DEFENSE, SECURITY AND STRATEGIES

THE F-22A FIGHTER JET

MODERNIZATION PROGRAM ANALYSES

DEFENSE, SECURITY AND STRATEGIES

Additional books in this series can be found on Nova's website under the Series tab.

Additional E-books in this series can be found on Nova's website under the E-book tab.

MILITARY AND VETERAN ISSUES

Additional books in this series can be found on Nova's website under the Series tab.

Additional E-books in this series can be found on Nova's website under the E-book tab.

DEFENSE, SECURITY AND STRATEGIES

THE F-22A FIGHTER JET

MODERNIZATION PROGRAM ANALYSES

GEORGE D. BORSARI
EDITOR

nova publishers
New York

For permission to use material from this book please contact us:
Telephone 631-231-7269; Fax 631-231-8175
Web Site: http://www.novapublishers.com

Library of Congress Cataloging-in-Publication Data

ISBN: 978-1-62417-326-4

Published by Nova Science Publishers, Inc. † New York

CONTENTS

PREFACE

The Air Force currently plans to spend $11.7 billion to modernize and improve reliability of the F-22A, its fifth generation air superiority fighter. Originally designed to counter air threats posed by the former Soviet Union, the post-Cold War era spurred efforts to add new missions and capabilities to the F-22A, including improved air-to-air and robust air-to-ground attack capabilities. This book provides an analysis of the modernization program for the F-22A fighter jet, with a focus on cost, technical and sustainment risks; and a comparison with the Legacy fighter modernization program.

Chapter 1 - GAO was asked to evaluate (1) cost and schedule outcomes and (2) testing results and risks going forward in the F-22A modernization program and related efforts. To do this, GAO examined the program's budgets and schedule estimates over time and discussed any changes with program officials, and reviewed progress and results from developmental and operational testing, and plans to mitigate risks and resolve system deficiencies.

Chapter 2 - The Air Force expects to invest a total of $9.7 billion in F-22A modernization through 2023. The Air Force and Navy have modernized many of their fighter and attack aircraft over the past several decades. Given this historical experience and concerns about the mounting cost of F-22A modernization, GAO was asked to examine the history of the modernization programs of the F-15, F-16, and F/A-18, and compare those legacy programs with the F-22A modernization program.

To identify differences and similarities between the F-22A modernization program and those of the selected legacy programs, GAO reviewed official service history documents and current and historical program documents; analyzed program cost, schedule, performance, and quantity data; and spoke with current and former Air Force, Navy, and contractor officials.

In: The F-22A Fighter Jet
Editor: George D. Borsari

ISBN: 978-1-62417-326-4

Chapter 1

TACTICAL AIRCRAFT: F-22A MODERNIZATION PROGRAM FACES COST, TECHNICAL, AND SUSTAINMENT RISKS*

United States Government Accountability Office

WHY GAO DID THIS STUDY

The Air Force currently plans to spend $11.7 billion to modernize and improve reliability of the F-22A, its fifth generation air superiority fighter. Originally designed to counter air threats posed by the former Soviet Union, the post-Cold War era spurred efforts to add new missions and capabilities to the F-22A, including improved air-to-air and robust air-toground attack capabilities. In 2003, the Air Force established the F-22A modernization program to develop and insert new capabilities in four increments.

GAO was asked to evaluate (1) cost and schedule outcomes and (2) testing results and risks going forward in the F-22A modernization program and related efforts. To do this, GAO examined the program's budgets and schedule estimates over time and discussed any changes with program officials, and

* This is an edited, reformatted and augmented version of United States Government Accountability Office, Publication No. GAO-12-447, dated May 2012.

reviewed progress and results from developmental and operational testing, and plans to mitigate risks and resolve system deficiencies.

What GAO Recommends

GAO recommends that DOD evaluate capabilities to determine if future F-22A modernization efforts meeting DOD policy and statutory requirements should be established as separate major acquisition programs.

DOD concurred with our recommendation.

What GAO Found

Total projected cost of the F-22A modernization program and related reliability and maintainability improvements more than doubled since the program started– from $5.4 billion to $11.7 billion–and the schedule for delivering full capabilities slipped 7 years, from 2010 to 2017. The content, scope, and phasing of planned capabilities also shifted over time with changes in requirements, priorities, and annual funding decisions. Visibility and oversight of the program's cost and schedule is hampered by a management structure that does not track and account for the full cost of specific capability increments. Substantial infrastructure costs for labs, testing, management, and other activities directly support modernization but are not charged to its projects. The Air Force plans to manage its fourth modernization increment as a separate major acquisition program, as defined in DOD policy and statutory requirements.

Testing of new capabilities to ensure operational effectiveness and suitability is ongoing. Results to date have been satisfactory but development and operational testing of the largest and most challenging sets of capabilities have not yet begun. Going forward, major challenges will be developing, integrating, and testing new hardware and software to counter emerging future threats. Other risks are associated with greater reliance on laboratory ground tests and relocating an F-22A lab needed to conduct software testing. While modernization is under way, the Air Force has undertaken parallel efforts to improve F-22A reliability and maintainability to ensure life-cycle sustainment of the fleet is affordable and to justify future modernization investments. But the fleet has not been able to meet a key reliability requirement, now changed, and operating and support costs are much greater than earlier estimated.

ABBREVIATIONS

AT&L	Acquisition, Technology and Logistics
DCMA	Defense Contract Management Agency
DOD	Department of Defense
DOT&E	Director, Operational Test and Evaluation
DTM	Directive-Type Memorandum
IFDL	Intra Flight Data Link
JDAM	Joint Direct Attack Munition
OSD	Office of the Secretary of Defense
O&S	Operating and Support
MADL	Multifunction Advanced Data Link
MDAP	Major Acquisition Defense Program
MTBM	Mean Time Between Maintenance
RAMMP	Reliability and Maintainability Maturation Program

The Honorable Daniel K. Inouye
Chairman
The Honorable Thad Cochran
Ranking Member
Subcommittee on Defense Committee on Appropriations
United States Senate

The Air Force currently plans to spend about $11.7 billion to modernize and improve the reliability of the F-22A, the Air Force's fifth generation air superiority fighter. About $9.7 billion will be spent on specific modernization increments and related support costs, and nearly $2 billion will be used to improve the reliability of the F-22A and make structural repairs. Originally designed to counter air threats posed by the former Soviet Union, the post-Cold War era spurred efforts to add new missions and capabilities to the F-22A, including improved air-to-air and robust airto-ground attack and capabilities. In 2003, the Air Force established the F-22A modernization program to develop and insert new capabilities. The timing and scope of the modernization program has changed over time, costs have significantly increased, and fielding of some capabilities has been delayed. In this context, you asked us to evaluate cost, schedule, and performance outcomes and risks of the F-22A modernization program.[1]

To determine the extent to which the F-22A modernization program is meeting cost and schedule goals, we researched the history of the program, plans and expectations at the start, and tracked budgets and schedule estimates over time. We identified changes in plans and estimates and discussed these changes with Air Force and Department of Defense (DOD) officials. To determine performance outcomes and risks remaining, we reviewed progress and results from developmental and operational testing and plans to mitigate risks and resolve system deficiencies. This included reviewing and discussing annual test reports from DOD's Office of the Director, Operational Test & Evaluation (DOT&E), briefings to defense oversight and requirements offices, summaries of recent operational test results provided by Air Force test officials, and program risk information related to testing new capabilities.

We conducted this performance audit from June 2011 to April 2012 in accordance with generally accepted government auditing standards. Those standards require that we plan and perform the audit to obtain sufficient, appropriate evidence to provide a reasonable basis for our findings and conclusions based on our audit objectives. We believe that the evidence obtained provides a reasonable basis for our findings and conclusions based on our audit objectives.

BACKGROUND

The Air Force's F-22A Raptor is the only operational fifth-generation tactical aircraft, incorporating a low observable (stealth) and highly maneuverable airframe, advanced integrated avionics, and a supercruise engine capable of sustained supersonic flight. The F-22A acquisition program began in 1991 with an intended development period of 12 years and a planned quantity of 648 aircraft. The system development and demonstration period eventually spanned 14 years, during which time threats, missions, and some requirements changed. In particular, the F-22A was originally designed to fly primarily air-to-air missions; however, since that time the Air Force has decided to add air-to-ground capabilities to the F-22A. Development costs substantially increased and total quantities were eventually decreased to 188 aircraft. When the final aircraft is delivered in May 2012, the F-22A acquisition program will be complete at a cost of $67.3 billion.

In 2003, the Air Force established a modernization program to develop and insert new and enhanced capabilities considered necessary to meet the threat.

According to Air Force officials, modernization is defined as a process of upgrading and modifying aircraft with a focus on adding new capabilities. The modernization is now proceeding in four related increments, each with multiple projects:

- Increment 2,[2] the initial phase of modernization, addressed some requirements deferred from the acquisition program and added some new ground attack capability. It has been fielded.
- Increment 3.1 began fielding in November 2011 and adds enhanced radar and enhanced air-to-ground attack capabilities.
- Increment 3.2A is a software upgrade to increase the F-22A's electronic protection, combat identification, and Link-16 communications and data link capabilities.
- Increment 3.2B will increase the F-22A's electronic protection, geo-location, and Intra Flight Data Link (IFDL) capabilities, and adds AIM-9X and AIM-120D missiles.

In addition to these efforts, in 2006, the Air Force began a Reliability and Maintainability Maturation Program (RAMMP). Although the Air Force does not consider this part of the modernization program, it is integral to making the F-22A weapon system more available, reliable, and maintainable.

Since the F-22A's initial fielding in 2006, maintenance issues have prevented it from achieving reliability and availability requirements, and fleet operating and support (O&S) costs are much higher than projected earlier in the program.

F-22A Modernization Costs Have Increased and Deliveries of New Capabilities to the Warfighter Have Been Delayed

Total projected cost of the F-22A modernization program has more than doubled since it started. While the program has completed and fielded some of its planned capabilities, the overall schedule to complete integration and testing of planned capabilities and deliver them to the warfighter has slipped by nearly 7 years. The content, scope, and phasing of planned capabilities also shifted over time with changes in requirements, priorities, and annual funding decisions.

Visibility and oversight of the program's cost and schedule is hampered by a management structure that does not directly track and account for the full cost of specific capability increments. The Air Force plans to separately break out and manage the fourth increment as a major defense acquisition program, which should improve management and oversight.

F-22A Modernization Costs Have Risen Sharply Since the Program Began

The Air Force is now expected to spend around $11.7 billion to modernize and improve the reliability of the F-22A, compared with the $5.4 billion projected soon after the start of development. Officials underestimated the scope of the total program and the time and money that would eventually be needed to develop and field new capabilities. Contributing factors to this cost growth include (1) changed and added requirements; (2) unexpected expenses for building a support infrastructure; and (3) unplanned efforts to improve aircraft reliability and maintainability.

Program officials also said that instability in modernization funding contributed to some of the cost growth by stretching the time required to complete projects. Figure 1 shows increased cost estimates over time for the modernization program and other related costs.

Modernization increments include development and procurement costs directly tied to one of the four increments for acquiring upgraded capabilities. These include infrastructure costs for lab support, test operations, program management, retrofit to bring all aircraft to a common configuration, and other efforts integral to supporting modernization increments. Other improvement costs principally include the RAMMP reliability and maintainability projects and making structural repairs needed for the aircraft to achieve its required 8,000 hour service life.

At this point, an estimated $5.5 billion of the $11.7 billon has been spent. A future investment of around $6.2 billion remains: $1.3 billion for Increment 3.2B, $3.6 billion for other modernization and support activities, and $1.3 billion for completing the RAMMP and structural repairs.

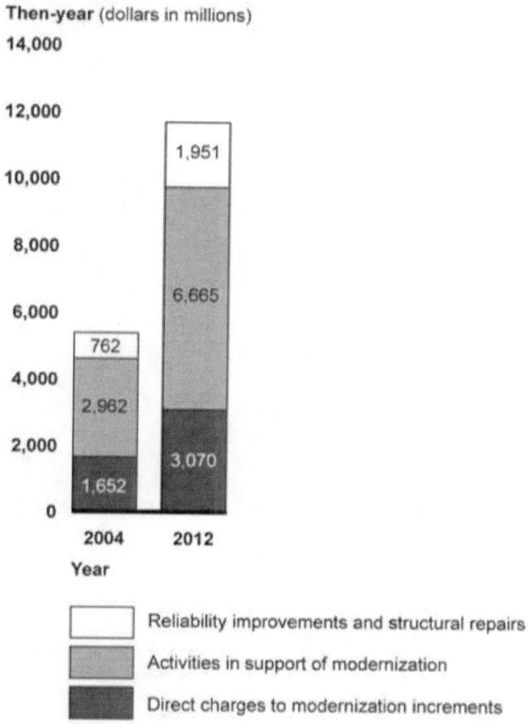

Source: GAO analysis of U.S. Air Force data.

Note: The 2004 estimate reflects projected costs from 2003 to 2012. The 2012 estimate reflects costs from 2003 to 2023.

Figure 1. Comparison of Estimated F-22A Modernization Program and Related Costs (Nominal Year Dollars in Millions, not Inflation Adjusted).

Completion of F-22A Modernization Projects Has Been Significantly Delayed as Content, Scope, and Phasing of Capabilities Changed over Time

When the F-22A modernization development program began, the Air Force expected to have all current planned capabilities integrated and fielding started by 2010. Now, the final increment is not expected to begin fielding until 2017, 7 years later than initially planned. Air Force officials stated that they underestimated the sheer magnitude of the modernization effort, both in the amount of time required to develop and integrate the capability, and costs to complete the modernization. According to program officials, contributing factors to delays include (1) additional requirements, (2) unexpected problems and delays during testing, and (3) research, development, testing, and evaluation funding fluctuations. Figure 2 compares the initial and latest schedules.

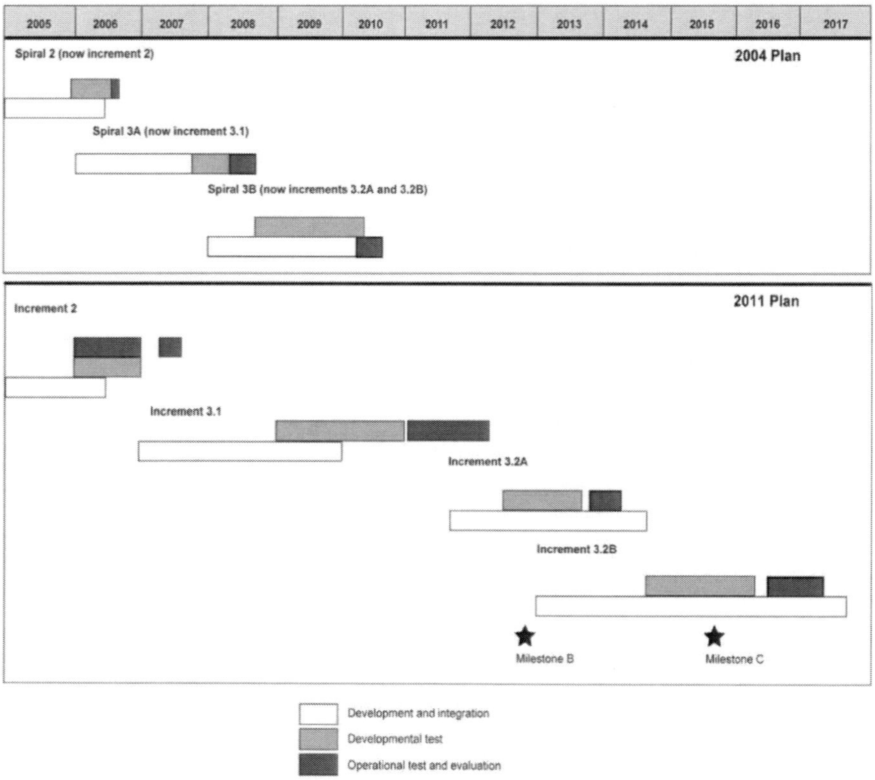

Source: GAO analysos of U.S. Air Force data.

Note: The Air Force has now replaced the term spiral development with increments which more accurately defines their approach to modernization. DOD Instruction 5000.02, Operation of the Defense Acquisition System (Dec. 8, 2008), provides for three decision points or phases. They include: milestone A (entry point for the technology development phase); milestone B (entry point for engineering and manufacturing development phase—which is comprised of two major efforts called integrated system design, and system capability and manufacturing process demonstration); and milestone C (entry point for the production and deployment phase).

Figure 2. Initial and Latest F-22A Modernization Schedules (as of December 2011).

According to Air Force officials, the program currently intends to upgrade 143 aircraft with the full complement of modernized capabilities by fiscal year 2020 and retain 36 aircraft with only Increment 2 capabilities to be used in training.

Increment 2	2005	2008	2011
Candidate 38A	●	●	●
Chaff Track Delete	●	●	●
Full Chaff Fix	●	●	●
Intra-Flight Data Link Widenet	●	●	●
Launch Zone Improvement	●	●	●
Over-the-Horizon Data Transmit	●	○	○
Supersonic JDAM	●	●	●
Increment 3.1	**2005**	**2008**	**2011**
Air to Ground Radar	●	●	●
Airborne Network Transmit	●	○	○
Electronic Attack	●	●	●
Geolocate 1	○	●	●
Geolocate 2	●	○	○
JDAM Self Target	●	●	●
SDB Basic	●	●	●
Increments 3.2A and 3.2B	**2005**	**2008**	**2011**
AIM-9X[b]	○	●	●
AIM-120D[b]	●	●	●
Automatic Ground Collision Avoidance System	●	●	○
Autonomous Search	●	○	○
Combat ID[a]	○	●	●
Electronic Protection Update[a&b]	●	●	●
Enhanced Data Recording	●	○	○
Geolocate 1.5	○	●	○
Geolocate 2[b]	○	○	●
Geolocate 3 and 3A	●	○	○
Ground Moving Target Indication and Tracking	●	○	○
IFDL Upgrades[b]	○	○	●
Multi-function Advanced Data Link	○	●	○
Multi-function Radar	●	○	○
Link 16 Upgrades[a]	○	●	●
SDB Full	●	●	○
Selective Available and Anti-Spoofing Module	●	○	○

● Included in increment as of this year

○ Not included in increment as of this year

Source: GAO analysos of U.S. Air Force data.

Note: Upgrades generally provide classified capabilities, communications, cockpit displays, radar, situational awareness, and weapons.

[a] Included in Increment 3.2A.

[b] Included in Increment 3.2B.

Figure 3. F-22A Modernization Planned Capabilities in Increments Have Changed over Time.

Increment 3.1 is being fielded in fiscal years 2011 to 2016 and Increment 3.2A from fiscal years 2014 to 2016. Increment 3.2B, the last currently planned increment, is expected to field from fiscal years 2017 and 2020. Future capability enhancements are expected to follow the current modernization program, but have not been defined.

The content, scope, and phasing plan changed over time, contributing to cost and schedule problems. Figure 3 illustrates the changing nature of modernization projects, particularly in the later increments.

Some capabilities, such as the Multifunction Advanced Data Link, have been eliminated because of changes in requirements and immature technology. Some, like the AIM-9X missile, have been added to the program to meet emerging threats. Some required capabilities have been reduced, such as the Geolocate project, which will now field a less-capable version than initially planned.

Air Force officials stated that potential new capabilities are analyzed and vetted by evaluating technical maturity and applying cost as independent variable principles[3] to determine which to include in the F-22A modernization program. As a result of this evaluation process, certain capabilities have been modified, deferred, added, or eliminated. Most changes affect the final two increments. For example, MADL, which was intended to provide communications interoperability with the F-35 Joint Strike Fighter, was removed from Increment 3.2B. MADL and other deferred efforts, such as the full Small Diameter Bomb capability, may eventually be delivered in future increments yet to be defined.

VISIBILITY AND OVERSIGHT OF THE PROGRAM'S COST AND SCHEDULE IS HAMPERED BY A MANAGEMENT STRUCTURE AND FUNDING MECHANISM

Tracking and accounting for the full and accurate cost of each modernization increment, and individual projects within each increment, are limited by the way the modernization program is structured, funded, and executed. As depicted in figure 4, only 26 percent of total projected costs can be traced directly to the four modernization increments. About 57 percent of total costs go to fund activities that support all the modernization efforts and the overall F-22A program but are not charged to specific increments. These activities include test operations, the building and use of government labs, management activities, retrofit efforts to bring the fleet to a common configuration, and other infrastructure accounts. The remaining 17 percent funds the RAMMP program and structural repairs. While Air Force officials do not consider these efforts as part of the funded modernization program, we note that these efforts are needed to improve fleet affordability and achieve desired aircraft life and thus integral to justify future modernization investments.

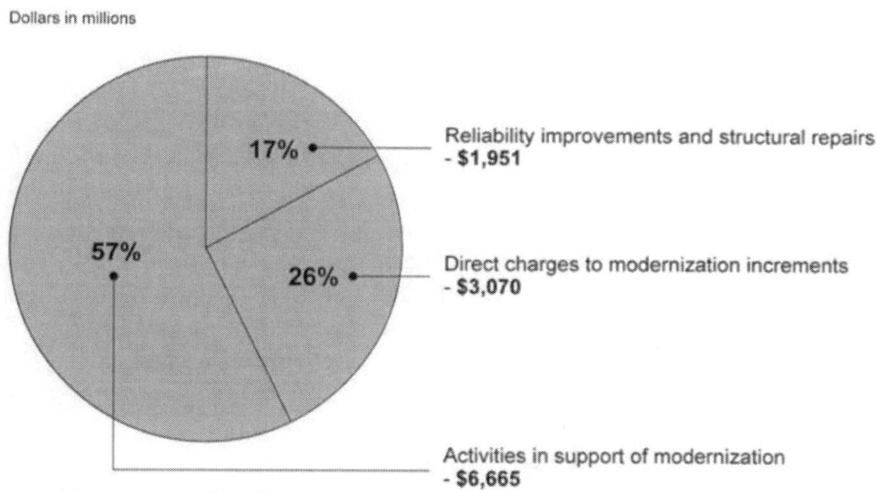

Source: GAO analysis of U.S. Air Force data.

Figure 4. Air Force Current Estimate of F-22A Modernization and Other Improvement Costs.

Program accountability and oversight have been hampered by how the modernization program was established, managed, and funded. As we reported in March 2005, the Air Force embarked on the modernization program without a knowledge-based business case to support the multibillion dollar investment to significantly change the aircraft's capabilities and missions.[4] We stated that the modernization program should have been established as an entirely separate acquisition program with a new business case because of the magnitude of the proposed changes. A sound business case would have matched requirements with resources—proven technologies, sufficient engineering capabilities, time, and funding—when undertaking new product development.[5] However, information about the schedule and funding was not adequately known at the start of modernization. Rather than making the new business case to justify and manage the modernization program as a separate major defense acquisition, Air Force officials incorporated it within the existing F-22A acquisition program and comingled funds. Their rationale was their belief that breaking modernization efforts out as a separate program would have delayed the capability. As a result, development funding and infrastructure expenses were added to the existing acquisition program's baseline.

The modernization program proceeded without establishing its own set of acquisition milestones and has not been subject to the same level of scrutiny by senior defense leaders or the performance reporting required of major defense acquisition programs as provided for in DOD acquisition policy. At their discretion, DOD chose to execute it within the baseline F-22A program.[6]

In November 2004, defense leaders recognized that the size and importance of the modernization program warranted a higher level of scrutiny. The acting Under Secretary of Defense for Acquisition, Technology and Logistics directed the Air Force to hold separate milestone reviews for the future stages of the modernization program to be consistent with DOD acquisition policy. Under this Air Force direction, the current modernization projects would not require formal milestones, but Office of the Secretary of Defense (OSD) oversight would be provided by periodic reviews. In 2007, OSD directed the Air Force to update the F-22A Acquisition Program Baseline to reflect the approved Increments 3.1 and 3.2; however, the Air Force believed that an acquisition strategy report rather than a baseline would provide better insight into funding and schedule details for the modernization increments. Not separating the modernization program from the F-22A program baseline was consistent with how the Air Force had handled modernization programs for prior aircraft. However, had the Air Force initiated the program under existing guidelines established by DOD Instruction 5000.02 for managing and implementing major acquisition programs, oversight of the program would have benefitted. Under these guidelines, programs are required to have an approved minimum set of Key Performance Parameters, included in the Capability Development Document; an approved Acquisition Strategy; Acquisition Program Baseline; an Analysis of Alternatives; and an Independent Cost Estimate, for a Milestone B decision that would allow them to proceed into the Engineering and Manufacturing Development phase.

OSD recently reiterated its requirement for the F-22A to be consistent with DOD policy, and in December 2011, OSD directed the Air Force to establish increment 3.2B as a separate major defense acquisition program. According to the Air Force, this increment is expected to cost around $1.5

billion. Given the significant slips in schedule experienced by increments 3.1 and 3.2A, the decision to separately oversee increment 3.2B is a late but positive change. Increment 3.2B will be reported as its own major program with system development starting in fiscal year 2013. This should improve management, cost visibility, and program oversight. Air Force officials told us that they expect to manage and report all future F-22A modernization programs as separate acquisitions, starting with Increment 3.2B.

PERFORMANCE OUTCOMES HAVE BEEN JUDGED SATISFACTORY, BUT TESTING AND IMPROVING RELIABILITY AND AFFORDABILITY OF THE FLEET WILL BE CHALLENGING

Testing how well new capabilities perform is ongoing; results to date have been satisfactory but development and operational testing of the largest and most challenging sets of capabilities have not yet begun. Going forward, major challenges will be developing, integrating, and testing new hardware and software to counter emerging future threats. Other risks are associated with availability of unique test assets, greater reliance on laboratory ground tests, and relocation of a key F-22A lab that is needed to help support testing of software for the new capabilities. Parallel efforts to improve F-22A reliability and maintainability are critical to ensure life-cycle sustainment of the fleet is affordable and to justify future modernization investments.

Operational Testing Results on New Capabilities Have Been Mostly Positive to Date, but More Challenging Efforts Are Still Ahead

New F-22A capabilities delivered by the modernization program will be demonstrated through follow-on operational testing and evaluation to assess the upgraded F-22A's effectiveness and suitability.[7] Testing on the first two increments successfully demonstrated new air-to-ground capabilities. Testing of the third and fourth increments has not begun and several technical risks remain for these new capabilities. Successful mitigation of these risks is critical to keeping F-22A's planned upgrades on schedule and within planned costs. Table 1 shows the current status of F-22A modernization operational testing for each increment.

Table 1. Current Status of F-22A Modernization Operational Testing for Each Increment, as of December 2011

Increment	Capabilities tested	Key issues during operational testing	Operational testing completed
Increment 2	Fixes to baseline aircraft and expanded air-to-ground capabilities against fixed targets with 1,000 pound Joint Direct Attack Munition.	Inspection and repair of low observable components required significant effort, and accounted for half of all maintenance hours.	August 2007
Increment 3.1	Additional air-to-ground capabilities allowing the F-22 to find and fix targets by itself, without the need for external platforms to provide coordinates.	Fleet stand-down due to concerns about the oxygen generation system; unavailability of a test range for flight testing and technical delays related to ground support equipment. Operational requirement for reliability was changed during testing.	November 2011
Increment 3.2A	Updates to electronic protection, combat identification and targeting capabilities.	Operational testing is expected to begin in late 2013.	Early 2014 (estimated)
Increment 3.2B	Improved strike capabilities with AIM-9X and AIM-120D missiles, and more advanced geolocation and electronic protection capabilities.	Operational testing is expected to begin in late 2016.	Mid-2017 (estimated)

Source: GAO analysis of DOD data.

Increment 2

Follow-on operational testing and evaluation for F-22A fighters incorporating Increment 2 capabilities, including assessments of expanded air-to-ground capability and improvements in system suitability, were successfully completed in August 2007. The F-22A's configured with Increment 2 capabilities were found to be operationally effective in suppressing and destroying fixed enemy air defenses, and also demonstrated successful fixes of deficiencies and weapons integration problems that had caused problems in previous testing. Flight testing demonstrated the ability to employ the Joint Direct Attack Munition (JDAM) at supersonic speeds in a high- threat anti-access environment where stealth capabilities are needed.

Without this capability, baseline aircraft were only able to launch JDAMs at fixed targets in lower threat environments and at slower speeds while using target coordinates from ground spotters.

Increment 3.1

Increment 3.1 further enhances F-22A's air-to-ground capability by allowing the aircraft to find and locate ground targets with on-board systems, rather than relying on external personnel and platforms for targeting. Increment 3.1 completed follow-on operational testing in November 2011; a significant delay of 4 years from the original plan due to shortcomings identified with the baseline and upgraded aircraft. In 2009 and 2010, the Director of Operational Test and Evaluation (DOT&E) reported significant stealth-related maintenance issues that lowered operational availability and mission capability rates. F-22A program officials identified technical issues in upgrading radar, navigation, and software that needed to be addressed to meet operational testing requirements.

The Air Force began Increment 3.1 operational testing in January 2011, but soon encountered flight delays that persisted from March to September 2011. The entire F-22A fleet was ordered to stand-down due to potential problems with the aircraft's oxygen generation system. Unavailability of the test range and technical problems with ground support equipment also contributed to the lengthy flight delay. The Air Force completed flight testing for Increment 3.1 in November 2011 and expects to release the operational test report in late March 2012. In its 2011 annual report, DOT&E did not identify any significant remaining issues since flights had resumed. DOT&E also approved reducing trials from 16 to 8 and decreasing simulator test trials from 96 to 64. According to program officials, hardware and software issues had been identified and fixed as testing progressed and test pilots provided very positive feedback on Increment 3.1's enhancements.

Increment 3.2A

Increment 3.2A development began in November 2011 after significant delays. This increment involves updating software to enhance electronic protection and combat identification capabilities, so that F-22A can handle new threats expected in the future. Developmental testing for this increment is expected to start in 2012 and be completed in late 2013. Operational testing and evaluation will follow and is planned to conclude in 2014.

Program officials assessed the Increment 3.2A schedule as having moderate risk. Test aircraft have been operating much longer than planned and were to be replaced by new production aircraft; however, this has not happened due to the substantial reduction in the size of the F-22A fleet. Other risks appear lower. For example, some software for electronic protection and combat identification capabilities has already been developed for the F-35 Joint Strike Fighter. Also, while the Link-16 upgrade will involve a significant amount of development work, program officials consider it to be moderate risk.

Increment 3.2B

Increment 3.2B is scheduled to begin Engineering and Manufacturing Development in December 2012 and the decision to enter into production is scheduled for January 2016. Key capability upgrades include integrating the AIM-9X and AIM-120D missiles on the F-22A and upgrading geolocation and electronic protection subsystems. Early requirements analysis determined that AIM-9X integration may be more difficult and take longer than expected and officials have already begun risk reduction efforts. Overall, software integration is considered to have the highest risk for Increment 3.2B projects, while hardware development is rated as a moderate risk. Program officials believe that the full range of capabilities added in the modernization program can be accommodated within the weight and space limitations of the F-22A aircraft, but this will be a critical consideration in any future modernization plans.

The Air Force is seeking ways to reduce the costs of Increments 3.2A and 3.2B by streamlining program activities. Officials want to make more use of developmental tests to also satisfy operational test requirements, allowing the program to identify errors for correction earlier and reducing overall costs by eliminating redundant tests. The program also intends to increase its use of F-22A ground laboratories to substitute for more expensive flight tests. The F-22A lab infrastructure is an extensive, distributed system of dedicated labs that integrate and certify flight software releases to the field and support F-22A modernization, production and sustainment activities. However, there are technical risks if lab tests do not fully replicate the performance of actual F-22A aircraft in intended environments. Officials are also expecting to save money by relocating the Raptor Avionics Integration Lab—a critical work site that stimulates sensors for targeting—from Marietta, Georgia, to Ogden Air Logistics Center, Utah by the summer of 2012. Program officials acknowledge there are some risks in this. For example, unique equipment could be damaged during the move and experienced lab staff could decide to leave the F-22A program rather than relocate.

RAMMP Program Addresses Reliability and Maintainability Deficiencies to Improve Affordability and Justify Future Investments

In addition to capability upgrades, the F-22A budget also funds efforts to address reliability and maintainability deficiencies that have increased support costs and have prevented the F-22A from meeting a key performance requirement. RAMMP is to develop and implement enhancements to increase aircraft availability, make maintenance faster and less costly, and reduce total life-cycle operating and support costs and cost per flying hour. While RAMMP is expected to reduce life-cycle costs over the long term, up-front investments to help realize future cost reductions have increased. The program had planned to spend about $258 million between 2005 and 2011, but actual investments through 2011 were about $528 million. The total RAMMP funding requirement through the year 2023 is now estimated at almost $1.3 billion. Air Force officials attributed part of RAMMP's increased costs to additional projects and increased labor hours to address corrosion.

Keeping the F-22A fleet affordable and meeting required performance measures is critical to sustaining fleet operations over the long term, and ensuring it is available in sufficient numbers for required missions. Projected operational and support costs are much higher than earlier estimates. For example, a 2007 independent estimate by the Air Force Cost Analysis Agency projected a $49,549 cost per flying hour in 2015 (by which time the F-22A was expected to reach full maturity), more than double the $23,282 cost per flight hour estimated in 2005.

Air Force officials gave various reasons for sustainment cost increases including (1) unrealized savings from the F-22A's performance-based logistics contract[8] (2) fixed costs that had to be spread over a smaller number of aircraft; and (3) higher than expected costs to refurbish or replace broken parts, including diminishing manufacturing sources. However, the one common contributing factor—and the most impactful— is the cost and complexity of maintaining stealth characteristics and restoring aircraft to the required stealth level after flight operations and maintenance. Our recent report found that the number of maintenance personnel required to maintain the F-22A's specialized stealth exterior has increased, posing a continuing support challenge for this aircraft.[9] This has important implications for the affordability and life-cycle cost estimates for the F-35 Joint Strike Fighter.

When it started in 2006, a major goal of RAMMP was to improve F-22A reliability to meet its key performance requirement by the time the fleet reached maturity at 100,000 total flight hours. This performance indicator, known as mean time between maintenance (MTBM), required aircraft in the F-22A fleet to fly an average of 3 hours between maintenance events, excluding routine servicing and inspections. This performance standard was a key performance requirement in the F-22A acquisition contract, but the fleet has never been able to meet that requirement. Currently, the MTBM achieved by the operational test aircraft with improvements is 2.47 hours.

In April 2011, the Joint Requirements Oversight Council approved changing the main reliability metric from MTBM to another performance indicator, known as material availability. Officials believed the MTBM indicator was hard to define and measure, was unrealistic, and did not accurately reflect the fleet's readiness to perform missions. Material availability is defined as the percentage of the fleet available to perform assigned missions at any given time. This standard calls for the F-22A fleet to achieve increasing levels of availability between 2011 and 2015 toward the final goal of 70.6 percent. Last year, the F-22A fleet achieved a 55.5 percent materiel availability rate. Stealth-related maintenance, system component reliability problems, and lack of spare engines were factors contributing to the fleet not achieving the goal. However, program officials expect the F-22A fleet to achieve the final availability goal by 2015 after the full fielding of reliability improvements. The Air Force reported that operational test on aircraft integrated with the current reliability improvements have achieved 78 percent availability; they anticipate significant gains by the overall fleet once reliability improvements are installed on all F-22A aircraft.

CONCLUSION

Keeping the F-22A as the world's most advanced stealth fighter requires the Air Force to counter changing threats, as well as ensure the F-22A fleet is affordable, reliable, and sustainable. In response to changing threats, officials began a Modernization Program to add new missions and capabilities while fixing problems and deficiencies that were carried over from the original development program. However, the F-22A modernization program has not had the management rigor or oversight on par with the $11.7 billion investment it entails. The program was not well-defined when it began in 2003, has had fluid scope and cost, and has been challenging from an oversight perspective as it was blended into the baseline F-22A program rather than being managed separately. As early as 2004, OSD began discussing the need to manage future modernization increments as separate acquisition programs. While modernization has been underway, the Air Force has found it necessary to invest in improved reliability and availability of the F-22A through the RAMMP program. The original reliability requirement was not met and has since been changed to another indicator. Meanwhile, O&S costs have been significantly higher than planned, with maintenance of the aircraft's stealth levels being particularly demanding. The lessons learned on the maintenance of the stealthy F-22A may have implications for the F-35 Joint Strike Fighter.

Splitting out increment 3.2B as a separate major acquisition defense program indicates that OSD is reasserting its role in the F-22A program. This is beneficial for oversight in light of the significant decisions and investments yet to come for the program. Increment 3.2B requires around $1.3 billion, while completing the RAMMP program, ongoing modernization projects, and other improvements will require an estimated $4.9 billion—a total future investment of around $6.2 billion. The program is highly dependent on a single contractor, whose responsibilities encompass managing the development and production of the F-22A; development, production, and retrofit of modernization; execution of the RAMMP program; and life-cycle support of the F-22A fleet, including supply and maintenance. Finally, the Air Force informed us that it expects to manage future modernization increments as separate acquisitions. However, given the approach the Air Force has taken to date on this and other modernization programs, there is little assurance that this will occur without specific OSD direction.

Recommendation for Executive Action

As new and enhanced capabilities are proposed and vetted beyond Increment 3.2B in the F-22A modernization program, we recommend that the Under Secretary for Acquisition, Technology and Logistics evaluate those capabilities in accordance with DOD policy and statutory criteria to determine if they should be established as separate major defense acquisition programs, each with its own milestones, business case, and cost baseline that includes all applicable direct and indirect support costs required to complete the program.

and data was somewhat constrained and our reporting limited to providing summary level observations due to the classification level of some of the data. Notwithstanding, DOD officials gave us access to sufficient information to make informed judgments on the matters covered in this report.

In performing our work, we obtained information and interviewed officials from the F-22A Program Office, Wright-Patterson Air Force Base, Ohio; Air Combat Command, Langley Air Force Base Virginia; Office of the Director, Operational Test & Evaluation, Office of the Secretary of Defense, Arlington, Virginia; and the Air Force Operational Test and Evaluation Center, Kirtland Air Force Base, New Mexico. We assessed the reliability of DOD and F-22A contractor data by (1) obtaining and reviewing related information from various sources, and (2) interviewing agency officials knowledgeable about the data. We determined that the data were sufficiently reliable for the purposes of this report. We conducted this performance audit from June 2011 to March 2012 in accordance with generally accepted government auditing standards. Those standards require that we plan and perform the audit to obtain sufficient, appropriate evidence to provide a reasonable basis for our findings and conclusions based on our audit objectives. We believe that the evidence obtained provides a reasonable basis for our findings and conclusions based on our audit objectives.

End Notes

[1] A companion report, GAO, *Tactical Aircraft: Comparison of F-22A Modernization Program and Legacy Fighter Modernization Programs,GAO-12-524* (Washington, D.C.: Apr. 26, 2012) addresses how the timing and strategy of the F-22A modernization program compares to similar past efforts on legacy fighter programs, including the Air Force's F-15 and F-16 and the Navy's F/A-18.

[2] The Air Force numbering scheme considers increment 1 to be the baseline capabilities delivered by the F-22A acquisition program.

[3] Cost as an independent variable principles call for the establishment of cost goals for operations, sustainment, and procurement, and for acquisition programs to make tradeoffs in terms of cost, schedule, and performance.

[4] GAO, *Tactical Aircraft: Air Force Still Needs Business Case to Support F/A-22 Quantities and Increased Capabilities*, GAO-05-304 (Washington, D.C.: Mar. 15, 2005).

[5] A business case is defined as demonstrated evidence that (1) the warfighter need exists and that it can best be met with the chosen concept and (2) the concept can be developed and produced within existing resources—including design knowledge, demonstrated technologies, adequate funding, and adequate time to deliver the product, GAO, *Defense Acquisitions: Managing Risk to Achieve Better Outcomes,* GAO-10-37T (Washington, D.C.: Jan. 20, 2010).

[6] Currently, a major defense acquisition program (MDAP) is a DOD acquisition program that is not a highly sensitive classified program and that is designated by the Under Secretary of Defense for AT&L as a MDAP or that is estimated to require an eventual total expenditure for research, development, test and evaluation, including all planned increments, of more than $365 million (based on fiscal year 2000 constant dollars) or an eventual total expenditure for procurement, including all planned increments, of more than $2.9 billion (based on fiscal year 2000 constant dollars). Directive-Type Memorandum (DTM) 09-027, *Implementation of the Weapon Systems Acquisition Reform Act of 2009* (Dec. 4, 2009, incorporating change 3, Dec. 9, 2011), Attachment 1 at § 13. See also, DOD Instruction 5000.02, *Operation of the Defense Acquisition System* (Dec. 8, 2008) and 10 U.S.C. § 2430.

[7] Operational effectiveness is the overall degree of mission accomplishment of a system when used by representative personnel in the environment planned or expected for operational employment and operational suitability is the degree to which a system can be satisfactorily placed in field use, with consideration given to reliability, availability, and maintainability.

[8] The F-22A's prime contractor, Lockheed-Martin, provides life-cycle product support, including supply and maintenance under this arrangement.

[9] GAO, *Defense Management: DOD Needs Better Information and Guidance to More Effectively Manage and Reduce Operating and Support Costs of Major Weapon Systems*, GAO 10-717 (Washington, D.C.: July 20, 2010).

In: The F-22A Fighter Jet
Editor: George D. Borsari

ISBN: 978-1-62417-326-4

Chapter 2

TACTICAL AIRCRAFT: COMPARISON OF F-22A AND LEGACY FIGHTER MODERNIZATION PROGRAMS*

United States Government Accountability Office

WHY GAO DID THIS STUDY

The Air Force expects to invest a total of $9.7 billion in F-22A modernization through 2023. The Air Force and Navy have modernized many of their fighter and attack aircraft over the past several decades. Given this historical experience and concerns about the mounting cost of F-22A modernization, GAO was asked to examine the history of the modernization programs of the F-15, F-16, and F/A-18, and compare those legacy programs with the F-22A modernization program.

To identify differences and similarities between the F-22A modernization program and those of the selected legacy programs, GAO reviewed official service history documents and current and historical program documents; analyzed program cost, schedule, performance, and quantity data; and spoke with current and former Air Force, Navy, and contractor officials.

DOD reviewed a draft of this report and had no formal written comments.

---

* This is an edited, reformatted and augmented version of United States Government Accountability Office, Publication No. GAO-12-524, dated April 2012.

WHAT GAO FOUND

The F-22A and legacy modernization programs GAO reviewed were rooted in different development strategies. The F-22A began as a single-step program and did not anticipate the need for future modernization, while the legacy programs each began with the expectation that their aircraft would be incrementally upgraded over time. F-22A modernization began in reaction to a major shift in the aircraft's basic mission, which required the development of new capabilities that had not been planned for as part of the initial development program. In contrast, the legacy modernization programs made planned incremental improvements to existing mission capabilities. All of the modernization programs began at about the same time in development and procurement. The F-22A program is developing and retrofitting new capabilities onto a complex stealth aircraft, which is costly—currently estimated at $9.7 billion total. Legacy modernization programs were less complex, and thus less costly, and incorporated mature technologies onto new production aircraft. Accurately identifying and comparing the total cost of each modernization program is difficult. Each of the programs, including the F-22A, initially managed and funded modernization as a continuation of its baseline program, so modernization costs and funding were not clearly identified in selected acquisition reports or budget documents.

ABBREVIATIONS

DOD	Department of Defense
HARM	High-speed Antiradiation Missile
IOC	initial operating capability
LANTIRN	Low Altitude Navigation and Targeting Infrared for Night
MSIP	Multinational Staged Improvement Program
OSD	Office of the Secretary of Defense
PEP-2000	Production Eagle Package-2000

The Honorable Daniel K. Inouye
Chairman

The Honorable Thad Cochran
Ranking Member

Subcommittee on Defense Committee on Appropriations
United States Senate

The Air Force currently expects to invest a total of $9.7 billion in its F-22A modernization program from 2003 through 2023.[1] Fighter aircraft modernization is not new within the Department of Defense (DOD). In fact, the Air Force and Navy have modernized many of their fighter and attack aircraft over the past several decades. Given this historical experience, and citing the mounting cost and timing of F-22A modernization, you requested that we examine the history of the modernization programs of the F-15, F-16, F/A-18, and F-117A, and compare those legacy programs with the F-22A modernization program. In response, this report identifies and discusses key differences and similarities in the F-22A modernization strategy and those of the legacy programs you identified.

To conduct our work we reviewed historical documents and data and discussed the development and modernization programs of the F-22A, as well as the F-15, F-16, F/A-18, and F-117A with current and former service and contractor officials. According to the service officials we spoke with, modernization is generally a process whereby upgrades and modifications are made in response to new requirements and to capitalize on advances in technology to increase an aircraft's capabilities over time. To better understand the overall development approach of each program, we reviewed selected acquisition reports [2], budget documents, program briefings, acquisition decision memorandums, official service history documents, prior GAO reports, and other pertinent documents. We met with, and in some cases reviewed the writings of, current and former Air Force, Navy, and contractor officials with firsthand knowledge of the programs we were reviewing in order to gain additional insights into when, why, and how they went about modernizing. In the course of our work, we talked with contractor and former program officials who provided limited insights into the initial development of the F-117A stealth aircraft. However, because of to the amount of time that had passed and the highly classified nature of the program, key documentation and data were not readily accessible. As a result, we were not able to collect sufficient information relevant to our reporting objectives and thus did not include the F-117A in this report. See appendix I for a more detailed discussion of our scope and methodology.

We conducted this performance audit from June 2011 to April 2012 in accordance with generally accepted government auditing standards. Those standards require that we plan and perform the audit to obtain sufficient, appropriate evidence to provide a reasonable basis for our findings and conclusions based on our audit objectives. We believe that the evidence obtained provides a reasonable basis for our findings and conclusions based on our audit objectives.

BACKGROUND

The Air Force's F-22A Raptor is the only operational tactical aircraft incorporating a low observable (stealth) and highly maneuverable airframe, advanced integrated avionics, and a supercruise engine capable of sustained supersonic flight. The F-22A acquisition program began in 1991 with an intended development period of 12 years and planned a procurement of 648 aircraft. The system development and demonstration period eventually spanned more than 14 years, during which time threats, missions, and a number of requirements changed. As a result, development costs substantially increased and procurement quantities greatly decreased—resulting in the procurement of only 179 aircraft, plus 9 development aircraft. The final aircraft is scheduled to be delivered in May 2012.

In 2003, the Air Force established a modernization program primarily to develop and insert new and enhanced ground attack capabilities that were considered necessary to meet current and future threats. The F-22A modernization program is broken into four phases, known as increments, with each phase being made up of multiple projects:

- Increment 2, the initial phase of modernization, addressed some requirements deferred from the acquisition program and added new ground attack capability.[3] It has been fielded.
- Increment 3.1 began fielding in November 2011 and adds enhanced radar and enhanced air-to-ground attack capabilities.
- Increment 3.2A is a software upgrade to increase the F-22A's electronic protection, combat identification, and capability to receive data over DOD's Link-16 data network.
- Increment 3.2B is expected to increase the F-22A's geolocation, electronic protection, and Intra Flight Data Link capabilities and integrate AIM-9X and AIM-120D missiles.

Some of the key content originally planned for Increment 3.2B has been deferred, the estimated cost of the overall modernization program has doubled, and the schedule has slipped by 7 years—which program officials attribute to requirements and funding instability. The most recent program schedule indicates that 3.2B will begin fielding in 2017.

F-22 AND LEGACY MODERNIZATION PROGRAMS TOOK DIFFERENT APPROACHES TO DEVELOPING AND FIELDING CAPABILITIES

The F-22A and legacy modernization programs we reviewed were rooted in very different development strategies, although they shared some similar characteristics. These differences and similarities are summarized in table 1.

Table 1. Comparison of the Differences and Similarities among Modernization Programs

Legacy aircraft	F-22A
Differences	
Initial development was incremental with plans to increase capabilities over time	Initial development was single step with no plans for future incremental upgrades
Initial development took 5 to 7 years	Initial development took 14 years
Ultimately procured thousands of aircraft	Ultimately procured 179 aircraft
Proactively modernized as requirements evolved and new technologies matured	Reactively modernized when a new mission was added
Incorporated upgrades into production lines and delivered new upgraded aircraft	Retrofitting upgrades into fielded aircraft because production has ended
Conventional aircraft designs and federated avionics reduced complexity and costs	Stealth aircraft design and integrated avionics make retrofits complex and costly
Similarities	
Began modernizing with more than 85 percent of estimated development costs funded	
Began modernizing with less than 33 percent of estimated procurement costs funded[a]	
Began modernizing with less than 20 percent of estimated procurement aircraft purchased[a]	
Managed modernization as a continuation of the original development program[b]	
Began modernizing while in production and around initial operating capability	

Source: GAO.

[a] The F-15 program was further into procurement when it began modernizing.

[b] Later increments of the F/A-18 program were funded and managed as distinct acquisition efforts, and the Air Force is planning to fund and manage F-22A Increment 3.2B as its own acquisition effort.

The F-22A began as a single-step program and did not anticipate the need for significant future modernization. The legacy programs, on the other hand, began with the expectation that their aircraft would be incrementally upgraded and modified over time. F-22A modernization began in reaction to a major shift in the aircraft's basic mission, which required the development of robust ground attack capabilities that were not part of the initial development program. In contrast, the legacy modernization programs were primarily initiated to make incremental improvements to existing mission capabilities. The F-22A and legacy modernization programs all began at about the same time in development and procurement. The F-22A program is more complex and costly than the legacy programs, primarily because the new capabilities have to be retrofitted onto complex, stealth aircraft that have integrated avionics systems, which, according to program and contractor officials, adds labor hours and cost. The current total estimated cost of F-22A modernization is $9.7 billion. The legacy programs incorporated planned incremental upgrades into new production aircraft that had less complex avionics systems and were not required to be stealthy. Because the legacy programs managed and funded modernization as a continuation of their baseline programs, it is difficult to isolate and compare the full costs of modernization.

Different Fundamental Development Strategies Laid Distinct Foundations for Modernization

The Air Force did not expect any major shifts in the F-22A's primary mission and thus did not plan for future modernization upgrades. From the outset, the Air Force's F-22A development strategy was to pursue a quantum leap in capability in a single development program, referred to as a single-step approach, to develop and field a stealthy aircraft with advanced capabilities to replace its aging F-15C/D fleet and perform air superiority missions. Recognizing the magnitude of this development effort, the Air Force estimated that it would need more than 12 years to develop and deliver an initial F-22A operating capability.

In 2003, we testified that the Air Force took on significant risk and onerous technological challenges by committing to an approach to F-22A development that promised to deliver all of the required capabilities in a single step.[4] We noted that while it may have allowed the F-22A program to compete for funding, it hamstrung the program with little knowledge about its true technology, funding, and schedule needs. In addition, the Air Force did not make early trade-offs between requirements and available resources and therefore never established an executable business case.[5] Ultimately F-22A development took more than 14 years, encountered significant cost increases and quantity reductions, and has not yet fully met established requirements, specifically those related to reliability and maintainability.

In contrast, the legacy programs we reviewed approached development of their aircraft as an incremental process in which initial capabilities would quickly be developed and fielded, and as requirements evolved and technologies became available, additional increments of aircraft would then be developed. All of the legacy programs encountered difficulties during their initial development programs, yet they all delivered initial increments of operational aircraft within 5 to 7 years. In some cases, the initial aircraft provided only a limited operating capability because certain performance requirements could not be achieved. In most cases, the users deemed the limited capability acceptable, with the understanding that future increments of the aircraft could address the shortfalls. The following are illustrative examples from each program[6] (see app. II for a more detailed discussion of each program):

- **F-15 Eagle:** In the late 1960s, the Air Force identified the need to develop and field a new air superiority aircraft to counter emerging threats. The new aircraft, designated the F-15 Eagle, was expected to possess advanced capabilities, excel in close combat and maneuvering situations, and specialize in the tactical missions including escort and combat air patrol. Prior to the start of full-scale development in 1970, key decision makers made trade-offs, such as opting not to include 2,000 pounds of additional fuel capacity, to ensure that the program would be able to deliver a useful increment of capability within a relatively short time frame. Those early trade-offs were made with the understanding that the traded capabilities could potentially be added in the future if requirements demanded them and resources were available. While the F-15 development program was considered expensive relative to other programs at that time, and encountered significant difficulties with its engines, it was able to deliver an initial operating capability in 1975—5 years after the start of development—and quickly fielded hundreds of operational aircraft.
- **F-16 Fighting Falcon:** The F-16 development program that began in 1975 was essentially the continuation of a competitive prototype program that had been ongoing since 1972 known as the Lightweight

Fighter Prototype program. In fact, the Lightweight Fighter Prototype program's requirements document became the initial basis of the F-16 full-scale development program. According to the requirements document, —the Air Force intend[ed] to investigate the feasibility and operational utility of highly maneuverable lightweight fighter aircraft through a prototype design, fabrication, and flight test program." In addition, the requirements emphasized the need for the aircraft to be —relatively low cost." The Air Force also expected the F-16 to be the low-end complement to its high-end F-15 fleet. The program was designated as a design-to-cost effort, meaning that cost was the key requirement against which all other requirements were traded. As a result, the baseline F-16 aircraft—ultimately designated as F-16A/B—were day-only, fair-weather fighters with relatively basic capabilities, although they did possess some more advanced capabilities like computer-aided flight controls known as fly-by-wire. The radar system in the initial F-16 aircraft did not fully meet performance specifications, and similar to the F-15, the aircraft had significant difficulties with its engine. Regardless, the Air Force delivered an initial operating capability just over 5 years after development start and quickly fielded hundreds of operational aircraft.

- **F/A-18 Hornet:** The Navy's F/A-18 development program that began in 1975 was rooted in the Lightweight Fighter Prototype program. Despite direction to procure the same aircraft as the Air Force—the F-16—the Navy chose to develop its own unique fighter and attack aircraft, citing the need for two engines and other unique features that it believed were necessary to operate from an aircraft carrier. As a result, the Navy developed its own requirements and planned for a 7-year development program that some documents indicate was also to be a design-to-cost program. According to current F/A-18 program officials, the Navy established a formal plan for future F/A-18 improvements and upgrades at the time the development program began. While the baseline F/A-18 program was able to achieve initial operating capability in 1983, slightly later than originally planned, it did not fully meet its established requirements for combat radius or —bring back" capacity—that is, the capacity of the aircraft to return to the aircraft carrier with unused weapons and fuel. Program officials pointed out, however, that those capability shortfalls were deemed acceptable by the warfighters, and deferred into the future to allow for the production and fielding of hundreds of operational F/A-18A/B aircraft.

A comparison of data from F-22A, F-15, F-16, and F/A-18 selected acquisition reports at 5-year intervals over the first 20 years following the start of development for each program shows that the incremental development approaches of the legacy systems quickly produced large quantities of operational aircraft and introduced several new increments of aircraft over that time span (see table 2). The data also highlight the high cost—in comparable 2012 dollars—of the overall F-22A program relative to the legacy aircraft we reviewed, and the relatively low quantities of baseline aircraft that its single-step development approach produced over that same time frame.

Table 2. Cumulative Cost, Quantity, and Capability Increments over 20 Years Following Development Start

Millions of 2012 dollars					
Program	Data type	Year 5	Year 10	Year 15	Year 20
F-22A	Cost	$16,605.1	$30,329.9	$56,830.0	$74,632.5
	Quantity (procured)	0	2	92	179
	Increments (series)	-	A	A	A
F-15	Cost	$12,961.7	$33,265.8	$44,042.3	$56,787.7
	Quantity (procured)	92	579	792	1,002
	Increments (series)	A/B	A/B, C/D	A/B, C/D	A/B, C/D, E
F-16	Cost	$10,109.6	$29,477.8	$51,315.9	$61,562.6
	Quantity (procured)	250	989	1,859	2,201
	Increments (blocks)	5, 10, 15	25, 30/32	25, 30/32, 40/42	25, 30/32, 40/42, 50/52
F/A-18	Cost	$5,788.3	$25,311.5	$41,750.9	$66,203.9[a]
	Quantity (procured)	9	325	745	979
	Increments (series)	A/B	A/B	A/B, C/D	A/B, C/D[a]

Source: GAO analysis of DOD data.

[a] Nearly $4 billion of the cumulative cost increase in the F/A-18 program between year 15 and year 20 is directly attributable to the beginning of F/A-18E/F development, but because the F/A-18E/F had not been fielded as of year 20 it is not listed as a new Increment (series) in this table.

Modernization Began for Different Reasons but at Similar Points in Development and Procurement

The F-22A modernization program began in reaction to a significant change in the aircraft's primary mission. In contrast, the legacy programs we reviewed established modernization requirements that focused on using mature technologies to upgrade the capabilities of their respective aircraft to better perform the missions for which they had been initially developed.[7] The F-22A and legacy programs we reviewed all began modernizing when their respective development programs were either complete or nearing completion—around the time that the baseline program achieved initial operating capability—and nearly all of the programs were early in procurement, with the exception of F-15, which was further along.

F-22A modernization was initiated in 2003, in response to new requirements for the aircraft to perform ground attack missions in addition to the air supremacy missions it had originally been designed for. Given the magnitude of this shift in mission, and because the original development program had not anticipated the need for such a change, critical information about requirements, technical scope, schedule, and funding was not available at the time modernization began. As a result, the initial cost and schedule estimates for the overall modernization program were not fully informed—that is, they were not knowledge based—and have since changed significantly, with costs doubling and schedule slipping by more than 7 years.

Because the legacy programs anticipated future upgrades, they began planning and working with other program offices, contractors, and in some cases, foreign governments, to identify potential new technologies for future increments while their initial development programs were ongoing. In many cases, this collaborative approach allowed programs to leverage investments in new technology that had been made by other programs or even foreign governments that had purchased variants of the respective programs' aircraft. The programs also worked closely with the warfighters to ensure that the technologies they were pursuing would provide new capabilities to address new requirements or would sufficiently enhance existing capabilities. In some cases, this approach required the warfighters to agree to eliminate or defer some desired capabilities indefinitely because they were not technologically feasible or because they were not affordable. For example, former F-16 program officials explained that the warfighters had stated a desire to include the Low Altitude Navigation and Targeting Infrared for Night (LANTIRN) subsystem in the first major F-16 upgrade that was expected to provide aircrew with the ability to fly both day and night and in adverse weather, while improving terrain-following and targeting capabilities. However, the technology was not mature at the time and as a result the program office deferred the capability. The former officials noted that the F-16 eventually received LANTIRN as part of its Block 40/42 upgrade, but only after the technology had been matured for use on the F-15E.

While making these types of trade-offs meant that not every desire would be met, it allowed the programs to establish sound business cases for moving forward with new increments, and provided the warfighters with some assurance that the end product would be delivered quickly and perform as expected. This type of proactive approach allowed the legacy programs to begin fielding their first increments of modernized aircraft—in each case designated as C/D series aircraft—in less than 5 years from the start of their respective modernization programs. The following examples from the F-16 and F/A-18 programs provide further illustration:

- **F-16 Fighting Falcon:** The Air Force established an F-16 Multinational Staged Improvement Program in February 1980. The program provided a structured means of incrementally modifying and upgrading the F-16, and was originally conceived with three stages. Although the specific content of each future stage was not fully defined at the outset, the program office worked closely with the warfighters, technology developers, and other program offices to establish feasible and affordable requirements as each successive stage approached. According to a RAND Corporation study[8] done for the Air Force in 1993, ―t[he staged improvement program was] essentially a management device for coordinating many concurrent efforts to integrate subsystems with one another and an F-16 airframe. That is, in each stage, new designs of the F-16 [were] conceived that integrate[d] many new subsystems to create a coherent aircraft with new combat capabilities." As new subsystems with potential for future F-16 integration were developed, the F-16 program office established a relationship with the subsystem program office, in some cases providing aircraft for testing and influencing design, to help mature the technology and facilitate future integration onto the F-16. In most cases, new technologies were incorporated onto new production aircraft on the production line. The first F-16C/D aircraft was delivered to the Air Force on schedule in December 1984, less than 5 years after the beginning of the improvement program.[9]

- **F/A-18 Hornet:** The first major upgrade of the F/A-18 began in 1984, with the issuance of an engineering change proposal. The Navy's stated goal was to improve the F/A-18's existing capabilities while also providing some new capabilities in the areas of avionics, armament, and electronic warfare.[10] The upgrade was also expected to provide new mission computers with adequate speed, interface, and memory capacity to facilitate future growth. The Navy planned to use technologies, or subsystems, that had been developed and matured outside of the F/A-18 program. In its December 1985 selected acquisition report to the Congress, the Navy noted that the new subsystems, or technologies, for the F/A-18 upgrade would be provided as government-furnished equipment. Therefore, the costs and risks associated with developing and maturing the technologies were not borne by the F/A-18 program. The program integrated the new technologies into its production line in 1986 and received the first F/A-18C/D aircraft as scheduled the following year—only 3 years after the engineering change proposal was first issued.

Although the programs began for different reasons, our analysis of program data provided to the Congress in selected acquisition reports in December of the year immediately preceding the start of each respective modernization program indicates that they began at essentially the same points in development and procurement. For example, we found that the F-16, F/A-18, and F-22A had all funded more than 85 percent of their projected development costs, funded less than one-third of their estimated procurement costs, and procured less than 20 percent of their total estimated aircraft quantities. The F-15 program was also nearly done with development when it started modernizing, but it was further along in procurement, having funded and procured more than 40 percent of its projected aircraft. Figure 1 compares and contrasts the specific percentages for each program, and to provide additional context, it identifies the total estimated costs—converted to 2012 dollars—and total procurement quantities from which the percentages were calculated.

All of the modernization programs also began around similar acquisition events. Our review of program selected acquisition reports and other program data found that all of the programs began modernizing after they had initiated production and around the time they achieved initial operating capability. It is important to note, however, that those acquisition events took place much later in the F-22A program, primarily because the original development program

was over 14 years long, while the legacy development programs were 5 to 7 years long.

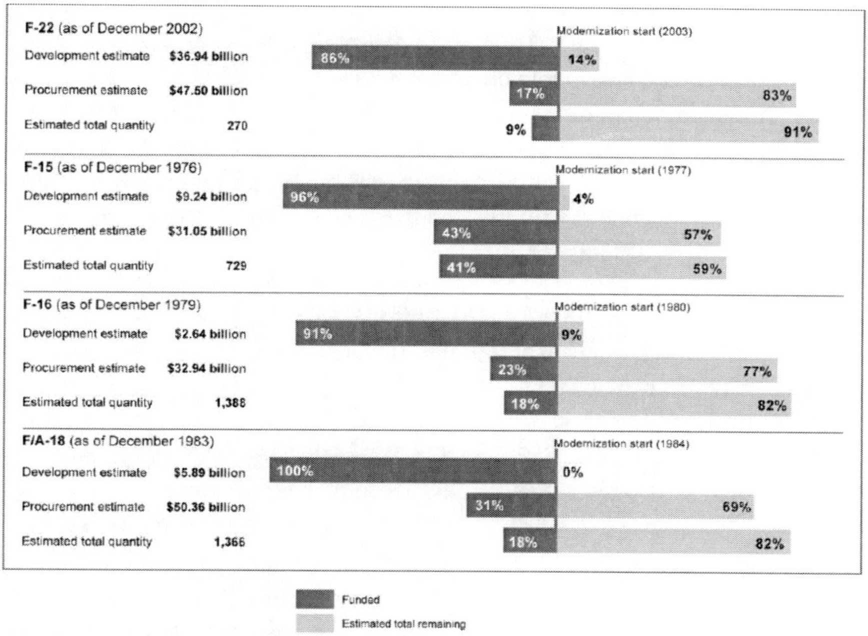

Source: GAO analysis of DOD data.

Figure 1. Comparison of Investment Progress Prior to Modernization.

F-22A Upgrades Are Being Retrofitted onto Complex, Stealth Aircraft While Legacy Programs Incorporated Upgrades into Production of New, Less Complex Aircraft

Nearly all F-22A modernization upgrades will have to be retrofitted onto fielded aircraft while the legacy programs integrated their upgrades into new production aircraft. The Air Force began integrating F-22A Increment 2 onto production aircraft in 2007, and received the first Increment 2 aircraft from the contractor the following year. All of the remaining aircraft were produced and delivered with Increment 2 upgrades incorporated. However, F-22A production was terminated in 2009, before the second modernization increment (Increment 3.1) had finished development, so the remaining modernization increments will have to be retrofitted into the fleet. As a result, the aircraft will have used up some of their service life by the time they are fully upgraded. Based on F-22A flight hour data provided by the program office our analysis indicates that a large number of aircraft are likely to have flown more than 1,500 hours, or nearly 20 percent of their 8,000-hour service lives, before the Increment 3.2B upgrades are fielded.[11] In contrast, the legacy programs produced entirely new upgraded aircraft.

It should also be noted that retrofitting upgrades onto stealth aircraft with fully integrated computer systems—referred to as fused or integrated avionics—like the F-22A is a riskier and more complex process than integrating new technologies into a conventional aircraft with separate and distinct computer systems and software for each subsystem—known as federated avionics—even if the technologies are mature. In large part, this is because any changes made to the F-22A have to conform to the aircraft's overall stealth design and will require updates to the aircraft's computer operating software. For example, the F-22A requires missiles that are carried and launched from internal weapons bays—not hung on the wings or under the fuselage as is the case with the F-16, F-15, and F/A-18. In addition, any new weapon added to the aircraft will also require new software to target and launch the weapon. For the F-22A, that software will have to be fully integrated into the aircraft's overall avionics system and tested thoroughly to determine its impact on all of the aircraft's other systems, which is costly and time consuming. Office of the Secretary of Defense officials point out that the legacy systems also had to integrate and test new software to ensure that it would work properly, but because the avionics systems were federated that process did not take as long and was less costly. In addition, the legacy aircraft had greater freedom to place new technologies onto the exterior of the aircraft or make structural changes as needed.

Given the stealth and avionics complexities of the F-22A, it is likely that it will be more costly to modernize than the F-15, F-16, or F/A-18. F-22A contractor officials emphasize that these complexities translate into labor hours and ultimately costs that the legacy programs would not have incurred. The total cost of F-22A modernization—through Increment 3.2B—is currently estimated to be $9.7 billion. We were not able to isolate comparable costs for the legacy modernization programs, primarily because they all funded and managed modernization as a continuation of their initial baselines. As a result, their respective selected acquisition reports and budget requests did not explicitly identify all of their modernization costs or funding needs.[12]

CONCLUDING OBSERVATIONS

As DOD pursues more complex and costly fighter aircraft to meet the advanced threats of the future, it is increasingly important that programs begin planning for incremental modernization at the earliest possible point. An overall incremental approach to development and early modernization planning were keys to the success of the legacy aircraft modernization programs we reviewed. From the beginning, they worked with the warfighters, technology developers, and in some cases foreign governments, to match requirements with available resources, and quickly developed and delivered new increments of upgraded aircraft to the warfighters. In most cases, the programs delivered initial increments of aircraft with limited capabilities, with the expectation that they would be upgraded over time as funding and technologies became available. Although the legacy and F-22A programs began modernizing at the same general points in time, the F-22A did not originally plan for a major modernization program, so when the aircraft's mission changed in 2003, the resources—primarily technology and funding—needed to meet the new requirements had not been fully developed or identified. As a result, the cost, schedule, and performance projections for the F-22A modernization program were not well founded and, over time, costs have doubled and the delivery of the full required capability has been delayed by more than 7 years. In addition, the majority of the F-22A modernization upgrades will be retrofitted onto fielded aircraft—a complex and costly undertaking—and by the time all of the required capabilities are fielded the amount of useful life remaining on the aircraft will likely be limited.

AGENCY COMMENTS

DOD reviewed a draft of this report and had no formal written comments. However, DOD did provide technical comments that were incorporated as appropriate.

Michael J. Sullivan, Director
Acquisition and Source Management

APPENDIX I: SCOPE AND METHODOLOGY

In order to compare and contrast the F-22A modernization program with those of the F-15, F-16, and F/A-18, we examined key program requirements and acquisition documents and prior GAO work. In addition, we interviewed knowledgeable Office of the Secretary of Defense (OSD), Air Force, Navy, and contractor officials, as well as former program officials. We also reviewed relevant studies done by the RAND Corporation and discussed those studies with knowledgeable RAND Corporation officials.

We obtained documents, data, and other information from officials at the Aeronautical Systems Center, Wright-Patterson Air Force Base, Ohio; Air Combat Command, Langley Air Force Base, Virginia; Naval Air Systems Command, Naval Air Station Patuxent River, Maryland; and Lockheed Martin Corporation. We were also asked to review the F-117A program, and while we were able to discuss the program with some former contractor and program officials, we were not able to collect sufficient information relevant to the objectives of this report.

The officials we met with noted that most of the primary source documentation related to the original program is either still highly classified or difficult to access because of the amount of time that has passed.

In conducting our analysis, we identified relevant cost, schedule, and requirements information from selected acquisition reports, budget documents, program briefings, and official service histories. To ensure that our comparisons reflected programs at common points in development and procurement, we first summarized and compared the data for each program at 5-, 10-, 15-, and 20-year intervals following the start of the initial development program.

We then used the start of modernization as our common point and identified and compared the cost, schedule, and quantity status of each program—in percentage terms to normalize the data—based on data in the selected acquisition reports provided to the Congress in December of the year preceding the start of modernization. According to the service officials we spoke with, modernization is generally a process whereby upgrades and modifications are made in response to new requirements and to capitalize on advances in technology to increase an aircraft's capabilities over time.

To estimate the likely number of flights hours that F-22A aircraft will have flown before Increment 3.2B is fielded, we obtained actual flight hour data from the F-22A program office for January 2007 through January 2012. We then calculated the average flight hours per year for each of those aircraft over that 5-year period, and used those averages to project the number of hours that each aircraft might fly from January 2012 through January 2018, around the time Increment 3.2B begins fielding.

We then added those projected hours to the actual number of hours flown by each aircraft through January 2012.

To assess the reliability of the program cost, funding, schedule, quantity, and flight hour data we used, we talked to agency officials about the processes and practices used to generate the data. We also corroborated the data by reviewing relevant documentation from various sources. We determined that the data were sufficiently reliable for the purposes of this report.

We conducted this performance audit from June 2011 to April 2012 in accordance with generally accepted government auditing standards. Those standards require that we plan and perform the audit to obtain sufficient, appropriate evidence to provide a reasonable basis for our findings and conclusions based on our audit objectives.

We believe that the evidence obtained provides a reasonable basis for our findings and conclusions based on our audit objectives.

APPENDIX II: SYSTEM MODERNIZATION SUMMARIES

This appendix provides more details on the modernization programs of the fighter and attack aircraft addressed in the body of this report. Each system summary includes a general overview of the modernization program and a more detailed discussion of key aspects of each modernization increment. Each program summary also includes a modernization timeline.

The timelines depict the length of the original development program and illustrate the amount of time between the start of development—represented by year 0—and other key program events, including initial operating capability (IOC), the beginning of modernization, and the delivery of new upgraded capabilities.

The F-22A timeline depicts the total estimated time frame for retrofitting the fleet with full global strike capability, from Increment 2 through Increment 3.2B, and the timelines for the F-15, F-16, and F/A-18 depict the time frames for the fielding of each new increment (series) of aircraft.

Source: U.S. Air Force.

Figure 2. F-22A Raptor.

Development start: June 1991
Initial operating capability: December 2005
Development cycle time: 14 years
Production cycle time: 11 years
Total aircraft procured: 179

Program Modernization Overview

In 2003, the Air Force established the F-22A modernization program in response to requirements for robust air-to-ground and other new capabilities. Around that same time, the Air Force also initiated efforts to improve the reliability and maintainability of the aircraft, although those efforts are not officially part of the modernization program. The new requirements represented a significant change from the F-22A's original air superiority mission. Initial development work on modernization enhancements started in 2003 and was initially planned to extend over a 7-year period with fielding of the full increment of required capabilities expected to start by 2010 and cost more than $4 billion. However, the program has since been restructured—largely because of requirements changes and funding instability according to program officials. Program officials are not sure when the full required capability will be delivered, and the total amount invested from 2003 through 2023 is currently estimated to be $9.7 billion.

The Air Force plans to achieve the full increment of air-to-ground attack capability by developing portions of that capability and retrofitting them into its F-22A fleet in several phases. The first phase—known as Increment 2—has been fielded. The second phase—known as Increment 3.1—has completed operational testing and is now being retrofitted onto the aircraft. The Air Force expects to issue the final operational test and evaluation report for Increment 3.1 in 2012. The third phase—known as Increment 3.2—has been divided into two smaller phases referred to as Increments 3.2A and 3.2B. Increment 3.2A is almost exclusively software focused and is currently in development. The Air Force will manage Increment 3.2B as a separate major defense acquisition program and recently received approval from OSD to begin preparation for a Milestone B review in the first quarter of fiscal year 2013. Figure 3 provides a timeline of F-22A modernization highlighting key events such as the start of the original aircraft development program, the beginning of production, the achievement of IOC, the start of the modernization program, and the overall time frame for fully fielding new global strike capabilities from the beginning of Increment 2 retrofits through the planned completion of Increment 3.2B retrofits.

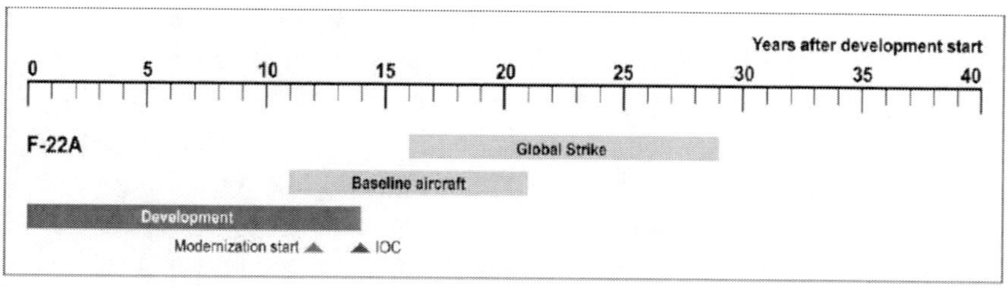

Source: GAO analysis of DOD data.

Figure 3. F-22A Modernization Timeline.

Increment 2

The Air Force began production of the F-22A before all of the basic program requirements had been met. As a result, it was forced to begin a modernization program to fully mature and retrofit technologies onto aircraft that had already been delivered to the war fighter. The Air Force began F-22A modernization in 2003 with Increment 2, which was the first of four phases needed to achieve the full required capability. This phase was expected to fix problems left over from the original development program and provide some basic air-to-ground attack capabilities. Many of the upgrades in this phase were software related, although the incorporation of the Joint Direct Attack Munition for basic air-to-ground missions was also included.

Follow-on operational testing and evaluation for F-22A fighters incorporating Increment 2 capabilities, including assessments of expanded air-to-ground mission capability and improvements in system suitability, were successfully completed in August 2007. The related upgrades were subsequently incorporated into the F-22A production line. Aircraft configured with Increment 2 capabilities were found to be operationally effective in suppressing and destroying fixed enemy air defenses, and also demonstrated fixes of some deficiencies and weapons integration problems that had been significant detractors in the original test program. Aircraft demonstrated the ability to employ the Joint Direct Attack Munition at supersonic speeds in a high-threat anti-access environment where stealth capabilities are needed. In comparison, the baseline aircraft (pre-Increment 2) were only capable of launching the Joint Direct Attack Munition at fixed targets in lower-threat environments and at slower speeds.

Increment 3.1

Increment 3.1 is now being fielded and adds enhanced air-to-ground attack and enhanced radar capabilities. These capabilities are expected to further enhance the F-22A's air-to-ground capability by allowing the aircraft to find and target ground targets with on-board systems, rather than relying on external personnel and platforms for targeting. Increment 3.1 began development in December 2006. The Air Force began retrofitting aircraft with Increment 3.1 capabilities in 2011 and expects to continue retrofitting through 2016. The Air Force began Increment 3.1 operational testing in January 2011 but soon encountered flight delays—because of the grounding of the fleet—that persisted from May to September 2011. However, the Air Force was able to complete flight testing for Increment 3.1 in November 2011 and now expects to release the operational test report in late March 2012.

Increment 3.2

Increment 3.2 has been broken into two phases, referred to as Increments 3.2A and 3.2B, and is expected to deliver additional advanced global strike capabilities. The Air Force initially expected this phase to deliver the final installment of capabilities that would meet the full air-toground requirements. The related capabilities include enhanced weapons, improved communications, and self-protection upgrades. The program office originally expected to begin fielding these capabilities in 2010, but according to program officials, requirements and funding instability have caused the program schedule to slip by more than 7 years, and they do not currently known when the full capability will be achieved. Increment 3.2A development began in November 2011 and is expected to update existing software to enhance Electronic Protection and Combat Identification capabilities. The first developmental test events for this increment are expected to start in late 2012 and be completed in late 2013. Operational testing is planned to conclude in early 2014 with fielding of Increment 3.2A planned to occur between 2014 and 2016. In December 2011, the Air Force received approval from OSD to begin the process of structuring Increment 3.2B as a new major defense acquisition program. The formal start of system development—Milestone B—on Increment 3.2B is planned for the first quarter of fiscal year 2013, with fielding to take place between 2017 and 2020. Key efforts in Increment 3.2B include integration of the AIM-9X and AIM-120D missiles and upgrading Geolocation and Electronic Protection subsystems.

Source: U.S. Air Force, Staff Sgt. Aaron Allmon.

Figure 4. F-15 Eagle/Strike Eagle.

Development start: January 1970
Initial operating capability: September 1975
Development cycle time: 5 years
Production cycle time: 30 years
Total aircraft procured: 1,074

Program Modernization Overview

For more than three decades, the Air Force has focused on upgrading and modifying the F-15 by defining, developing, and producing increments of militarily useful capabilities. The primary drivers behind F-15 modernization included basic capability improvements as well as the need to address capability gaps and respond to evolving threats. Since the beginning of the original development program in 1970, there have been two major F-15 upgrades: one that resulted in the production of the F-15C/D series aircraft and another more extensive upgrade that resulted in the production of the F-15E multimission aircraft series. Figure 5 provides a timeline of F-15 modernization highlighting several key program events, such as the start of the original aircraft development program, the beginning of production, the achievement of IOC, and the start of each major modernization upgrade. The figure also indicates when the first upgraded F-15C/D and F-15E were delivered to the Air Force.

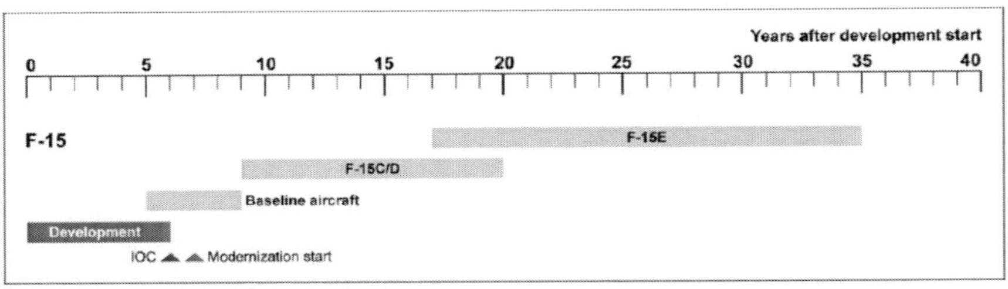

Source: GAO analysis of DOD data.

Figure 5. F-15 Modernization Timeline.

The Air Force began the F-15 program with the intent to quickly develop and acquire a weapon system capable of operating as an advanced, high-performance, air superiority fighter. The Air Force made several other changes to the F-15 platform to reduce costs and bring the program within funding constraints while still retaining an acceptable level of capability. In one instance, officials decided to reduce the size of the F-15's internal fuel tanks to reduce cost, which subsequently resulted in a reduction in the aircraft's mission radius. Officials believed that a reduced internal fuel capacity would still provide them with an acceptable mission radius, along with the added benefit of reducing the size, weight, and cost of the aircraft. Several changes were also made to the F-15's avionics systems to make the aircraft more affordable. Although the F-15 program ended up being costly relative to other programs at the time, the Air Force was able to begin delivering initial increments of operational aircraft within 5 years.

Although the F-15 was able to develop and deliver an initial operating capability of its baseline aircraft (F-15A/B) just 5 years after development, as well as quickly deliver new increments of F-15C/D and F-15E aircraft, the overall program was not without difficulties. For example, in the early years the F-15A/B program had problems with its engines as well as performance problems with its tactical electronic warfare system. Additionally, the F-15C/D experienced problems with malfunctioning landing gear and the F-15E had difficulties with electronic warfare software development.

F-15C/D

The transition from F-15A/B[1] to F-15C/D represented the first significant F-15 modernization effort. By the mid 1970s the Air Force had determined that the baseline F-15 needed to be upgraded, largely to increase the aircraft's mission radius, which had been reduced earlier in the program to achieve cost savings. This first major F-15 upgrade was managed and funded as a continuation of the baseline F-15 program. As such, management and investment decisions were primarily made at the service and program levels and did not go through the higher-level— OSD—review and approval process. The Air Force did not develop or produce any detailed mission needs statement or requirements documentation specifically for the upgrade program. No distinct cost or schedule baselines were developed, funding was requested and provided through the F-15 baseline budget, and program progress was reported to the Congress through the existing F-15 selected acquisition reports. The Air Force's approach to upgrading the F-15 focused on incorporating new technologies into a modified variant of the basic F-15 airframe.

The Air Force recognized that the F-15 upgrade, which was designated Production Eagle Package-2000 (PEP-2000), was going to require significant modifications to the aircraft's airframe. The primary focus of PEP-2000 was to increase the aircraft's internal fuel capacity by 2,000 pounds, thereby increasing mission radius. The upgrade was authorized in October 1976 and development began in January 1977, just 16 months after the first F-15 squadron achieved initial operating capability. In addition to adding 2,000 pounds of fuel, the PEP-2000 program was also focused on incorporating

8,000 pounds of additional equipment, such as provisions for carrying exterior conformal fuel pallets, and improved landing gear. The F-15A with PEP-2000 would become the F-15C, and the F-15B with PEP-2000 would become the F-15D. The first F-15C was delivered to the Air Force in May 1979, just a little over 2 years after the PEP-2000 program began.

F-15E

In 1982, the Air Force began to evaluate the need for a multirole (air-to-air and air-to-ground) fighter capable of operating at night and in adverse weather conditions. The Air Force received competitive demonstrations of advanced versions of the F-15 and F-16—designated the F-15E and F-16XL—from the respective aircraft contractors and on February 24, 1984, chose the F-15E to meet the dual-role fighter requirement. At the time the Air Force selected the F-15E to meet its multirole fighter requirement, the baseline F-15 had been operational for over 8 years and the F-15 program had produced 834 total aircraft.

Source: www.af.mil.

Figure 6. F-16 Fighting Falcon.

Development Start: April 1975
Initial Operating Capability: October 1980
Development Cycle Time: 5.5 years
Production Cycle Time: 28 years
Total Aircraft Procured: 2,231

The F-15E was designed to provide a long-range, large-payload capability to strike second echelon targets at night and in adverse weather while retaining superior air defense capability. Additionally, the F-15E, which was originally designed as a derivative of the F-15D two-seater, would support two crew members. While the F-15E was treated as a new aircraft build, it was managed and funded as a continuation of the baseline F-15 program, similar to the PEP-2000 upgrade. However, the F-15E was developed and procured under a new contract that required competition, while the PEP-2000 upgrade was developed and procured as a program management directive that required no competition and no new contract. Much like earlier F-15 models, the F-15E program was committed to using existing technologies as much as possible and chose to build on a mature and proven airframe, the F-15C/D. The Air Force received its first F-15E in March 1987, and the new aircraft achieved limited initial operating capability in September 1989.[2]

Program Modernization Overview

The Air Force has modernized the F-16 over the past three decades by making incremental upgrades and modifications to a baseline aircraft.

Over that time span, the Air Force has developed and fielded two major F-16 increments, or aircraft series, the F-16A/B and the F-16C/D.[3] Within those major increments several subgroups, or blocks of aircraft with common capabilities, have also been developed and produced. The first three blocks, Blocks 5, 10, and 15, were all F-16A/B series aircraft, while the last four blocks, Blocks 25, 30/32, 40/42, and 50/52, were all F-16C/D series aircraft.

In February 1980, several months before the baseline aircraft achieved initial operating capability, the Air Force established the Multinational Staged Improvement Program (MSIP) for the F-16, primarily to provide a structured means of incrementally modifying and upgrading the aircraft over time. The program was originally conceived with three stages, although the detailed content of each stage was not fully defined. Figure 7 provides a timeline of F-16 modernization highlighting key events, such as the start of the original aircraft development program, the beginning of production, the achievement of initial operating capability, and the start of the overall modernization program. The figure also indicates when the first upgraded F-16C/D was delivered to the Air Force and notes the start of each block upgrade.

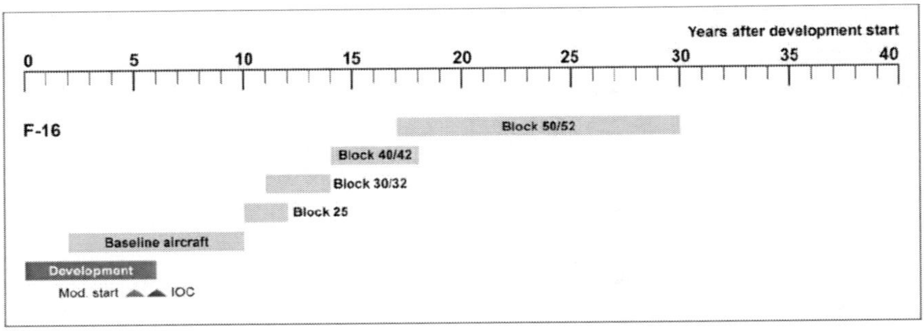

Source: GAO analysis of DOD data.

Figure 7. F-16 Modernization Timeline.

The Air Force's approach to the original F-16 development program provides insight into some of the basic management principles and practices that have continued to guide the aircraft's evolution. Those principles and practices have focused on ensuring affordability and technical feasibility before making large investments. This has required a consistent willingness on the part of decision makers and warfighters to accept incremental capability improvements and resist the pressure to attempt to make large capability leaps in a single step. This simplified, limited-capability approach has been evident from the beginning of F-16 development. When full-scale development of the F-16 began in April 1975, the Air Force had not documented any official requirements for the aircraft. Instead, the Air Force chose to use a concept paper issued in January 1973 to guide a fighter prototype program, known as the Lightweight Fighter Prototype Program. According to the concept paper, —The [lightweight fighter] prototype [was] expected to demonstrate in hardware the technology leading to a relatively low cost, high thrust-toweight ratio in advance of stated operational requirements." Given this early development work, and despite concerns about high levels of concurrency between development and production, the Air Force initiated F-16 production less than 2 years after the start of full-scale development and achieved initial operating capability in 1980, only 5 years after the start of development.

While the F-16 was able to quickly develop and deliver differing versions of the aircraft to the warfighter over time, it was not without some difficulty. For example, the F-16A/B had problems with its radar that required additional development and testing. The F-16C/D also experienced radar problems stemming from marginal performance and inaccurate readings.

Stage I: F-16A/B Block 15

The Air Force started F-16 modernization very early in the aircraft's life cycle, formally starting MSIP Stage I in February 1980. MSIP was funded and managed as a continuation of the original F-16 development program. This first stage primarily focused on making structural, wiring, interface, and cooling modifications, and resulted in the production of F-16A/B Block 15 aircraft. The modifications were relatively minor and did not require much new design or development work. However, the Air Force believed these modifications were essential to support future upgrades and preclude the need for costly modifications and retrofits. Also, as part of Stage I, the Air Force increased the size of the aircraft's vertical tail to address performance issues identified during flight testing. The aircraft contractor delivered the first Block 15 aircraft in November 1981, less than 2 years after Stage I began.

The Air Force worked closely with the aircraft contractor to identify and agree to the specific modifications that would be made as part of Stage I. Once agreement was reached, the Air Force and contractor had agreed to a number of structural, wiring, and interface modifications. However, the modifications required little new design or development work, so the Air Force was able to incorporate the changes directly onto the F-16 production line and quickly integrate them into production aircraft. In total, the Air Force expected to incorporate Stage I modifications into 455 production aircraft starting in November 1981. The RAND Corporation reported in 1993 that ―because little design work was required, developers viewed this stage as presenting little technical risk; rather, the main risk was associated with [Stage I] was whether provision made for future systems were the right ones. If future needs differed from those anticipated during Stage I, rework would be required to retrofit future systems."

Stage II: F-16C/D Blocks 25 and 30/32

The beginning of MSIP Stage II, which started 15 months after the beginning of Stage I, was officially authorized by the F-16 Joint Multinational Configuration Control Board in May 1981. Many of the Stage II modifications built on the provisions made during Stage I and focused on further increasing the aircraft's capacity to accommodate additional upgrades. One RAND Corporation official we met with told us that at one point early in the F-16 program the aircraft was ―gaining a pound a day" because of the number of requirements changes and related modifications that were being made. Stage II modifications were much more extensive than those in Stage I. The Stage II modifications included the incorporation of a new fire-control radar, additional electrical power and cooling capacity, additional computer memory, provisions for the Advanced Medium Range Air-to-Air Missile, and an alternate aircraft engine. The Air Force not only assigned Stage II aircraft unique block numbers—beginning with Block 25 and eventually moving to Block 30/32—but it also changed the aircraft series designation from F-16A/B to F-16C/D. The aircraft contractor delivered the first F-16C/D, a Block 25 aircraft, in December 1984, less than 4 years after Stage II began.

The Air Force's approach to MSIP Stage II was not significantly different from its approach to Stage I—that is, an incremental, but highly concurrent approach. Most of the modifications and upgrades planned for Stage II were based on variants of subsystems and technologies that were already in use on other fighter aircraft, bomber aircraft, or both. According to the RAND Corporation's detailed review of MSIP, the MSIP managers viewed the risk associated with this stage as low to moderate because of the stage's evolutionary nature. They pointed out that as new capabilities became available—that is, as technologies matured—the Air Force would begin to plan for the integration of those capabilities into the F-16 production line. Similar to the first MSIP stage, nearly all of the modifications and upgrades for Stage II were ultimately worked into the F-16 production line, with some limited retrofitting of fielded aircraft. The Air Force was able to work the changes into the production line largely because production of F-16A/B aircraft had continued during the early phases of Stage II. This highly concurrent environment continued. While the Block 25 aircraft were in production, the MSIP officials continued to explore additional possible modifications and ultimately made changes that resulted in another aircraft configuration, designated the Block 30/32. Between Blocks 25 and 30/32 the Air Force added new weapon capabilities and incorporated an alternate fighter engine.[4]

Stage III: F-16C/D Blocks 40/42 and 50/52

The Air Force received approval to start MSIP Stage III in June 1985. This stage introduced further advances in the F-16 fire-control radar and computer capacity, in addition to introducing night vision infrared navigation and targeting, global positioning system, and High-speed Antiradiation

Missiles (HARM), among other changes in weapon, radar, and avionics systems. Stage III ultimately resulted in two new F-16C/D block configurations—Block 40/42 and Block 50/52—each with its own distinct mission focus. According to the RAND Corporation, the Air Force's approach was a —deelopment-and-integration approach" that was marked by —the continual introduction of pre-planned changes and updates." During Stage III, the Air Force continued to monitor the technology market and work closely with technology developers to mature and integrate new capabilities into the aircraft while concurrently producing and fielding of Block 30/32 aircraft.

Source: www.navy.mil.

Figure 8. F/A-18 Hornet/Super Hornet/Growler.

Development start: December 1975
Initial operating capability: March 1983
Development cycle time: 7 years
Production cycle time: > 35 years
Total aircraft procured: 1,685

The key focus of the Block 40/42 modifications was the incorporation of nighttime flying and targeting capability, primarily provided by the Low Altitude Navigation and Targeting Infrared for Night (LANTIRN) system, to support precision strike missions, which had been deferred from the Block 25 program. In contrast, the Block 50/52 modifications focused on fully integrating the HARM targeting system to provide the capability to suppress enemy air defenses. While the development programs for both LANTIRN and HARM were managed and funded as distinct acquisition efforts, the F-16 program office provided aircraft for flight testing. This collaborative approach, which had also been used during previous MSIP stages, allowed the technology programs to mature while also allowing the F-16 program to address integration and performance problems before investing in significant modifications. The Air Force procured a combined total of 913 Block 40/42 and 50/52 aircraft between fiscal years 1987 and 2001. The first Block 40/42 aircraft was funded for procurement in fiscal year 1987 and delivered to the Air Force in December 1988—less than 4 years after Stage III began. The first Block 50/52 was funded for procurement in fiscal year 1990 and delivered to the Air Force in October 1991, over 6 years after Stage III began.

Program Modernization Overview

The Navy has taken an evolutionary approach to modernizing its F/A-18 aircraft through incremental upgrades and modifications. The Navy has focused on upgrading and modifying the F/A-18 by defining, developing, and producing increments of militarily useful capabilities—both improving existing capabilities and adding new capabilities. Since the beginning of the original development program in 1975, there have been three major F/A-18 upgrades that produced the F/A-18C/D; the F/A-18E/F, designated the Super Hornet; and the EA-18G, designated the Growler. Figure 9 provides a the timeline of F/A-18 modernization highlighting key events, such as the start of the original aircraft development program, the beginning of production, the achievement of initial operating capability, and the start of each major modernization upgrade. The figure also indicates when the first upgraded F/A-18C/D and F/A-18E/F aircraft were delivered to the Navy.

F/A-18 modernization has been driven by a combination of factors, including the need for regular capability improvements, the need to replace aging legacy aircraft to ensure that adequate force structure is maintained, and the need to respond to evolving threats and capability gaps with new available technologies. Program officials note that Navy has taken a basic evolutionary approach to modernizing the F/A-18. While the ultimate required capabilities—that is, the E/F or G series capabilities—had not been fully defined when the C/D series upgrades began, the engineering change proposal that guided the C/D series upgrade fully defined the requirements for that increment of capability.

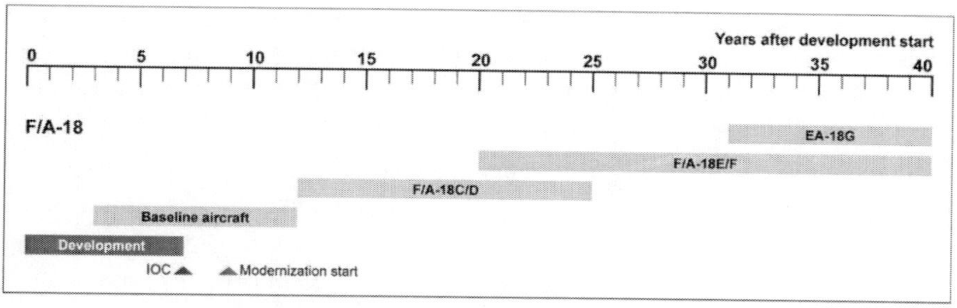

Source: GAO analysis of DOD data.

Figure 9. F/A-18 Modernization Timeline.

F/A-18C/D

The Navy initiated the first major F/A-18 upgrade relatively early in the program's life cycle with the issuance of an engineering change proposal in August 1984. At the time, the Navy had finished development of the F/A-18A/B[5] and was in the early stages of producing and fielding operational aircraft. The upgrade began just over 1 year after the first F/A-18 squadron achieved initial operating capability, just over 3 years after production started, and less than 9 years after the start of full-scale development. The decision to upgrade the F/A-18A/B was largely driven by the Navy's desire to keep the aircraft's capabilities current by taking advantage of newer and more advanced technologies. The Navy's primary goal in issuing the engineering change proposal was to improve existing capabilities and provide some new capabilities in the areas of avionics, armament, and electronic warfare.[6] In addition, the upgrade was expected to provide new mission computers with adequate speed, interface, and memory capacity to facilitate future growth. The Navy planned to use technologies, or subsystems, that had been developed and matured outside of the F/A-18 program. The program's December 1985 selected acquisition report noted that the technologies were expected to be provided to the program as government-furnished equipment. The engineering change proposal noted that the modifications would be made to aircraft on the production line, beginning with the aircraft planned for procurement in fiscal year 1986. The first F/A-18C/D aircraft was delivered to the Navy in October 1987, a little more than 3 years after modernization began. The Navy procured a total of 627 F/A-18C/D aircraft over a nearly 13-year period, with the final aircraft being delivered in August 2000.

The Navy managed and funded this first major F/A-18 upgrade as a continuation of the baseline program. As such, management and investment decisions were not required to go through higher-level DOD reviews, but were instead made at the service level by Navy leadership. The Navy did not develop or produce any detailed mission needs statement or requirements documentation to support the upgrade. No distinct cost or schedule baselines were developed, funding was requested and provided through the F/A-18 baseline budget, and program progress was reported to the Congress through the existing F/A-18 selected acquisition reports. Program officials note that the Navy's approach emphasized use of mature technologies and the integration of those technologies onto F/A-18 airframes that were already in production. They point out that with the exception of some minor modifications, the upgrade did not require any structural changes to the existing F/A-18 airframe.

F/A-18E/F

In May 1992, the Navy sought and received OSD approval to begin full-scale development of its second major F/A-18 upgrade, designated the F/A-18E/F Super Hornet.[7] At the time, the F/A-18 aircraft had been operational for nearly 10 years. The decision to develop the F/A-18E/F was based on the Navy's determination that it needed an upgraded carrier-based, multi-role fighter to ensure that it had enough operational aircraft to adequately man its aircraft carriers in the 1990s and early 2000s. The Navy expected the upgrade to address shortfalls in the F/A-18's range and ability to return to the carrier with unused weapons and stores (referred to as carrier recovery payload). In addition, the upgraded aircraft were expected to be stealthier and thus more survivable than the F/A-18A-D aircraft they would replace.

Unlike the first major upgrade, which was funded and managed as a continuation of the baseline program, the F/A-18E/F upgrade was approached as a new formal acquisition program. Although the Navy considered the F/A-18E/F development a modification—in its estimation, it was a logical continuation of the general F/A-18 upgrade strategy—the program was funded and managed as a new acquisition program within the framework of DOD's acquisition policies. The F/A-18E/F program was committed to using existing technologies as much as possible and chose to build on a mature and proven aircraft, the F/A-18C/D. The Navy received its first production F/A-18E in

December 1998, less than 7 years after development began, and achieved initial operating capability in September 2001. As of December 2010, the Navy expected to procure a total of 556 F/A-18E/F aircraft, with the final procurement projected to occur in fiscal year 2014.

EA-18G

The most recent variant of the F/A-18 aircraft is the EA-18G Growler, which is an electronic warfare-equipped F/A-18F. An analysis of alternatives was conducted to identify platforms that would be dedicated to providing advanced jamming capabilities for the suppression of enemy air defenses. The analysis was driven by a projected shortfall in the electronic attack inventory primarily caused by attrition and the increasing cost of operating the current aging fleet of EA-6B Prowlers. Based on the analysis of alternatives, the Navy chose to modify an F/A-18F instead of building an entirely new aircraft. OSD approved the start of EA-18G development in December 2003.

The Navy has managed the EA-18G program as a distinct major acquisition effort from the beginning. In fact, F/A-18 program officials emphasize that the EA-18G did not evolve from any F/A-18 modernization or preplanned improvement program, but instead was selected from an analysis of alternatives to replace critical airborne electronic attack capabilities. They also emphasize that the EA-18G has distinctly different mission—electronic warfare—and therefore, while it is a variant of the F/A-18F, it is not viewed as a new increment.

At the time the EA-18G development program started, the Navy still had a significant amount F/A-18E/F procurement remaining with approximately 370 aircraft still to be purchased, although it had already achieved initial operating capability. As a result, EA-18G development and procurement ended up being highly concurrent with F/A-18E/F procurement. The Navy received approval to begin EA-18G production in July 2007 and the aircraft achieved initial operating capability in September 2009. As of December 2010, the Navy anticipates procuring a total of 114 EA-18G aircraft, with the final procurement being made in fiscal year 2013.

End Notes

[1] In addition, the Air Force also expects to invest nearly $2 billion in F-22A reliability improvements from 2003 through 2023.

[2] Selected acquisition reports are key recurring summary status reports to the Congress on the cost, schedule, and performance of DOD's major defense acquisition programs. 10 U.S.C. § 2432.

[3] The Air Force numbering scheme considers Increment 1 to be the baseline capabilities delivered by the F-22A acquisition program.

[4] GAO, *Best Practices: Better Acquisition Outcomes Are Possible If DOD Can Apply Lessons from F/A-22 Program*, GAO-03-645T (Washington, D.C.: Apr. 11, 2003).

[5] Based on GAO's past work, a business case is defined as demonstrated evidence that the warfighter need exists and that it can best be met with the chosen concept and the concept can be developed and produced within existing resources—including design knowledge, demonstrated technologies, adequate funding, and adequate time to deliver the product.

[6] In addition to the F-15, F-16, and F/A-18 programs, we were also asked to look at the F-117A program, but because sufficient historical documentation and data were not readily accessible it is not included here as an example.

[7] The F-15E Strike Eagle and the EA-18G Growler are two cases where upgrades were driven by significant mission changes, similar to the F-22A. However, unlike the F-22A, those new F-15 and F/A-18 variants benefitted from knowledge gained through the operational use of hundreds of previously fielded aircraft.

[8] RAND Corporation, *The F-16 Multinational Staged Improvement Program: A Case Study of Risk Assessment and Risk Management*, N-3619-AF (Santa Monica, Calif.: 1993).

[9] Although the F-16 modernization program began in February 1980, the first stage of the program simply focused on adding structural and wiring provisions to the baseline aircraft to support future upgrades. The aircraft receiving those upgrades were ultimately designated as Block 15 aircraft. The first major upgrade came in the second stage of modernization, leading to development and delivery of the first F-16C/D series aircraft, designated as a Block 25.

[10] These capability upgrades were primarily expected to come from the addition of the Advanced Medium Range Air-to-Air Missile, Joint Tactical Information Distribution System, Flight Incident Recorder and Aircraft Monitoring System, and Advanced Self Protection Jammer.

[11] We obtained actual flight hour data from the F-22A program office for January 2007 through January 2012. We then calculated the average flight hours per year for each of those aircraft over that 5-year period, and used those averages to project the number of hours that each aircraft might fly from January 2012 through January 2018, around the time Increment 3.2B begins fielding. We then added those projected hours to the actual number of hours flown by each aircraft through January 2012.

[12] The F/A-18E/F and EA-18G programs are two exceptions. Both of these later F/A-18 upgrades were essentially managed as new acquisition efforts, and thus had distinct cost and funding baselines, selected acquisition reports, and budget requests.

End Notes for Appendix II

[1] The primary distinction between F-15As and F-15Bs is that A-series aircraft have a single seat cockpit while the B-series aircraft have a dual-seat cockpit and are used for training. The same distinction exists between F-15C and F-15D.

[2] The F-15E achieved a limited initial operating capability on September 30, 1989. The limited operational capability resulted from the lack of certain features, including automatic terrain following, the Low Altitude Navigation and Targeting Infrared for Night pod, a new ammunition feed system, and in part from delayed installation of the electronic countermeasures system.

[3] The A and B series aircraft are essentially identical with the exception of the number of seats in the cockpit. The A series aircraft have one seat while the B series aircraft have two. This same distinction exists for the C and D series aircraft as well. Given this high degree of commonality, each series pairing is typically referred to as a single unified increment, like A/B and C/D and not four distinct increments.

[4] The difference between the Block 30 and Block 32 aircraft is simply which fighter engine the aircraft have. The Air Force carried this same distinction forward into the Block 40/42 and 50/52 aircraft as well.

[5] The primary distinction between F/A-18As and F/A-18Bs is that A-series aircraft have a single-seat cockpit while the B-series aircraft have a dual-seat cockpit. The same distinction exists between F/A-18C and F/A-18D, as well as the F/A-18E and F/A-18F series aircraft.

[6] These capability upgrades were primarily expected to come from the addition of the Advanced Medium Range Air-to-Air Missile, Joint Tactical Information Distribution System, Flight Incident Recorder and Aircraft Monitoring System, and Advanced Self Protection Jammer.

[7] In January 1988, more than 4 years before beginning full-scale development of the F/A-18E/F, the Navy and the aircraft contractor began to study concepts for a more advanced variant of the F/A-18, which was referred to as Hornet 2000. At the same time, the Navy awarded a fixed-price development contract for the A-12 stealth aircraft, which subsequently encountered significant cost and schedule problems, and was ultimately terminated by the Secretary of Defense in 1991.

INDEX

F

G

H

I

L

M